BRITISH RAILWAYS STINKS

BRITISH RAILWAYS STINKS

The Life and Work of Britain's Last Railway Chemists

David Smith and colleagues

GRESLEY
BOOKS

British Railways Stinks is a light-hearted, accurate look at the work of the last
generation of railway chemists told in their own words. It would not have been
possible without the invaluable contributions of the following co-authors:

Geoff Hunt
Ian McEwen
Ian Cotter
John Sheldon
Vince Morris

Vince was our editor and played a major role in the production of this book,
and unfortunately did not live to see the fruit of his labours. This book is
dedicated to him and all the railway chemists who went before.

Published in Great Britain in 2019
by Gresley Books
an imprint of Mortons Books Ltd.
Media Centre
Morton Way
Horncastle LN9 6JR
www.mortonsbooks.co.uk

Copyright © Gresley Books, 2019

ISBN 978 1 911658 26 9

Typeset by Kelvin Clements
Printed and bound by Gutenberg Press, Malta

Acknowledgements:
The authors would like to acknowledge the help each has given to the others, and the input from
other colleagues who have supplied memories of their time as a railway stink. We must also thank
our inspiring leaders through the years – the Heads of Scientific Services – for their patience and
good-hearted responses to our antics, particularly Stan Bairstow, Eric Henley, Jim Ward and Granville
Morley. They were young once, apparently. Jim valiantly gave up a gentle snooze on a train from Derby
to Newcastle to review our first efforts. His observations and comments did not result in a full re-write;
it just seemed that way. He subsequently went caravanning in Wales to spend his time reviewing
the final text while the rain came down, and ensured the book was completed in a timely fashion.

Our thanks to:
Vicki Nye, who produced the caricatures of all the authors. We have forgiven her.
Corrina Paraskeva for the cartoons, which capture some of the more bizarre moments and visions.
Roger Hughes and Bob Symcox for additional anecdotes.
Roger Hughes, Dale Green, Arne Bale and Chris Harrison for additional photographs.

Contents

WHIFF

Apologies are acknowledged to Rudyard Kipling

If you can make the vilest stinks invented
    And work in them from morn till late at night
Or with your lot be perfectly contented
    When you are asked to fool with dynamite;
If you can still remain quite calm and placid,
    While depot managers effervesce and fret;
Or being told to test a fuming acid,
    Can suck it through a five ml pipette

If you can subjugate all thoughts of pleasure,
    And still retain a little self-esteem;
If you can give your few short hours of leisure
    To keeping up with every modern theme;
If you'll donate your every waking minute
    And seek your sole reward in duty done...
Yours is the lab and everything that's in it
    And, what is more, you're welcome to it, son

Introduction

I F HISTORY was just a series of dates, it would be easy. The first railway chemical laboratory was opened in 1864 by the London & North Western Railway at Crewe, and the last ones lost their direct link to the rail industry on their privatisation in 1996. But history is not just dates; it is people too, and within those dates there were many characters of extraordinary ability. Luckily their exploits in the 19th and early 20th centuries have been recorded in the book *Early Railway Chemists* by Colin Russell and John Hudson (Royal Society of Chemistry, Cambridge 2012), and the organisational background is meticulously documented in a monograph by Alastair Gilchrist, a former Deputy Director of Research, dated June 2012 (private circulation). However, neither record aspects of daily life in laboratories and on site which allowed the chemists to make their indelible, but usually anonymous, mark on the railway industry.

Whatever their expertise, every railway chemist has been asked the same question: "What do you actually do?" The most appropriate answer was probably that given by Leonard Archbutt, the Midland Railway Chief Chemist who, in 1914, commented: "I used to sometimes be asked what a railway company could find for a chemist to do; the difficulty would be to answer in what direction of its numerous ramifications the railway company is unable to find work for the chemist. When I was first appointed, some 30 years ago, there were very few railway chemists; and even we ourselves, much less those appointing us, had any adequate idea of the work there was for us to do."

Another approach would be to describe some of the jobs tackled, and that is precisely what this book attempts to do. It is written by members of the last generation of railway chemists and tells, in no particular order, of their experiences from the mid-1950s to the 1990s. It is not a technical treatise - it is a light-hearted, accurate look at the day-to-day goings-on in one of the more obscure corners of the railway industry. It covers many aspects of the work, from a BR chemist going to San Francisco to blow up a water melon to declaring an empty coal wagon a confined space, from whitewashing an in-service passenger train in a couple of seconds to questioning (on chemical grounds) the mental state of the chairman of British Rail, from gassing weevils to setting fire to a canal in Derby... I could go on, but this is only the introduction. You will have to read the book to find out all the details.

There is one thing which separates railway chemists from their more illustrious colleagues who work exclusively with other chemists in research or development departments. The industry is dominated on the technical side by engineers and on the operating side by talented employees steeped in generations of ordinary people getting the job done. This meant that the railway chemist was regarded with curiosity, suspicion or awe; often the 'scientist' in a group of very non like-minded people. It was both humbling and a privilege to work with dedicated people from a variety of backgrounds, but with the common aim of keeping the trains running. We were, probably, regarded as the 'nuts' of the organisation but we like to think we made a difference.

During the period covered by our anecdotes, the world of health and safety changed beyond all recognition, and both our working practice and expertise reflect this. Our approach to safety in the early days may seem cavalier but was appropriate for the time. After the introduction of the Health & Safety at Work etc. Act in 1974, when the management's role became a matter of legal concern, our work in this area - and our own approach to it — changed dramatically and, we trust, proved invaluable to the whole industry. No chemists were hurt during the preparation of this book.

And the title? Most chemical laboratories in schools were called the Stink Room and at heart we were just grown-up schoolboys. As you will read, Scientific Services could stink out a whole town.

Enjoy – we did.

Vince Morris

CHAPTER I

A (Very) Brief History of Scientific Services

Chemistry has been involved in the railway industry from the very beginning, writes Vince Morris.

B URNING COAL, steam raising and lubricating moving parts all involve chemistry. Consulting chemists offered their services to the fledgling railway industry for everything from water quality to metal analysis, but it was not until 1864 that a railway company, the London & North Western Railway, employed its own in-house chemist, Mr E Swann, and opened a laboratory in Crewe for him to work from. The industry continued to employ chemists right up until its privatisation. On 9th December 1996, Mr Swann's successors ceased to be employed by British Rail and were transferred to a private non-railway company, Atesta Group Ltd, thus ending 132 years of the railway chemist.

First, a bit of history. If you really cannot stand it, skip to the next chapter. Still with me? Then carry on…

The development of railways in the early part of the 19th century not only had vast social implications; it radically changed several industries, not least those supplying the metal rails, wooden sleepers, stone, water and coal. The railway even changed market gardening, as it demanded vast numbers of hedges as barriers to trespass. In 1857 the first steel rails were laid in England, at Derby Station, replacing cast and wrought iron which suffered brittle failures. But the steel was of varying quality and it was clear that quality control was vital: this required analysis, and therefore required chemists. When London & North Western opened a chemical laboratory at Crewe — the 'railway town' it created in Cheshire — laboratory staff initially concentrated on analysing steel and water for locomotive boilers. Soon the usefulness of having chemists available meant they became involved in other work, such as damage claims against the company, transportation of dangerous goods and accident investigation. Other railway companies saw the advantage, and soon the larger ones were opening their own laboratories and expanding the work the chemist was asked to undertake.

In 1894 the goods managers of the various railways were faced with the problem of transporting ether, which, due to its volatile nature, can exert considerable pressure on the drums in which it is stored. They decided to ask the chemists of the Midland, Great Western and Great Eastern Railways to advise them on the most suitable containers to use. Reportedly they could not agree, so the chemists, joined by their colleague from the Great Northern Railway, met at the Railway Clearing House, the organisation which oversaw cross-company transport arrangements, to decide on a joint response to the question. Thus, the Chemists' Committee was formed, and expertise exchanged between the various companies for the benefit of the whole industry. Their view, it seemed, was that 'good science' was more important than competitive advantage.

At the grouping of the plethora of railway companies in 1923, fourteen of the constituent companies had their own chemist or chemistry department. All four companies formed at the grouping — LNER, LMS, GWR and SR — developed their own chemistry departments, whose main roles were still seen as quality control and ad hoc investigations into operational problems. In 1948 the 'Big Four' were nationalised to form British Railways (BR), itself a part of the Transport Executive. BR soon identified the need for greater research into railway-orientated subjects and in 1951 BR Research was formed, with separate divisions for chemistry, engineering, metallurgy and physics. The latter three divisions were more occupied with developing new ideas or techniques, including obtaining a patent for spaceship design! The chemists, however, had their feet more firmly on the ground, still dealing with quality control, day-to-day operation of the railway and troubleshooting but now with access to experts in other scientific, technical and engineering disciplines. Although the chemistry division had specialist laboratories dealing with protective coatings (i.e. paint), corrosion, textiles, infestation and building materials, as originally constituted, it was split into two in 1962: Chemical Research and Regional Scientific Services. The newly formed Scientific Services absorbed some of the tasks and staff of the specialist laboratories, allowing it to undertake a broader scope of work.

When the new laboratories — to be known as the Railway Technical Centre (RTC) and designed to bring the main research effort of all the divisions together — were completed on London Road, Derby, in the late 1960s, such inter-disciplinary access became even easier. One result of the progressive opening of the RTC was that laboratory space became available to allow the Derby Area Laboratory to move from its cramped accommodation in Calvert Street into Hartley House, the original LMS laboratory on London Road, opposite the RTC, previously occupied by the Engineering Research Division. At the same time, a single-storey extension was built onto Hartley House for the newly created Central Analytical Laboratory (CAL) of Scientific Services. This facility was enthusiastically described in the British Railways Board annual report for 1964 as being "provided with the most up-to-date equipment which enables analyses to be done much more quickly and cheaply and also provides more comprehensive information." Although headquartered in Derby, Scientific Services maintained its network of area laboratories, roughly corresponding to the old company territories, allowing far greater cooperation on the ground than could be achieved from a remote ivory tower within the intimidating sprawl of the RTC.

We are now up to the period covered by the anecdotes in the rest of this book: the 1950s to the 1990s. At the beginning of this period there were still some residual company laboratories, such as Darlington, Horwich, Stonebridge Park, Stratford, Wimbledon and Ashford; by the late 1960s, the network of area laboratories was established. These were based on existing facilities in the resoundingly familiar railway locations of Glasgow, Crewe, Doncaster, Derby and Swindon, plus a new London laboratory at Muswell Hill. Why, you may ask, was there a laboratory at Muswell Hill? The brave new world required a laboratory not hidebound by pre-nationalisation boundaries within the capital (would Paddington ever speak to Euston, let alone Waterloo?), able to cater for the Southern Region of BR as well as the London end of the other regions. Following the 1954 closure of the line from Finsbury Park to Alexandra Palace, the station site at the Palace end of the branch fell into disrepair but the land was deemed suitable to build the London laboratory, even though it no longer had a rail link. Alexandra Palace is perched at the top of a hill, a terminal moraine left over from the ice age, and the argument was that "railway men would beat a path up the hill to the doors of the laboratory". They didn't, and although the lab opened in 1959, mail and samples were forever sent via Kings Cross with a daily collection by the laboratory van. It is claimed that many visitors "beating a path to the laboratory" gave up, exhausted, confused and frustrated, and were never seen again.

Despite the name Scientific Services implying a wide range of disciplines, the laboratories were staffed almost exclusively by chemists, recognising that either chemists were such good scientists that they could turn their hand to anything, or that they were so unemployable they would accept any work to justify their existence. Recruits were issued with the 'bible' — otherwise known as the analysts' handbook; three loose-leaf volumes outlining every procedure that the authors expected staff to encounter in their career... wonderful and factually accurate but completely outdated, both in the operation of the railway and in developing chemical techniques. In fact, most new recruits spent their formative years in the oils or water laboratories or on depot visits. Occupational hygiene — ranging from asbestos identification and monitoring, to entry into confined spaces and measuring noise levels — was a growing area of interest, prompted by the Health & Safety at Work etc. Act of 1974, and considered well within the capabilities of the trained chemist.

The oil laboratories were responsible for ensuring that lubricating oils within the diesel fleets of British Rail were functioning satisfactorily. Covering several thousand miles a week and with 80-gallon sumps you did not, as with the car, change the oil every 10,000 miles, so the oil's viscosity and contaminant content were assessed every few days. Additionally, samples were examined for traces of metals or salts, the presence of which would indicate wear within the engine, allowing potential failures to be predicted in advance and corrective action to be taken before failure in service. Locomotives had their own health card to allow trends to be established.

By this time the steam locomotive was virtually extinct, and the old trade of checking boiler water for suitability was dying out. The water laboratory looked at the microbiology of drinking water samples obtained daily from restaurant cars, both to check the on-board sterilising equipment and to avoid passengers developing food poisoning. Depot visits were precisely that: a regular visit to depots to check their effluent discharges met consent conditions, that the chemicals about the site were correctly stored and used and that staff using them were suitably protected.

As well as these routine functions, there were the more esoteric requests from anywhere within the railway family for scientific input into problems, be it smells within high speed train (HST) coaches, premature failures of wiring in new coaches, fires in tamping machines, selecting the most reflective materials for high visibility clothing, suspected radiological leaks from nuclear flasks or frost-damaged potatoes.

Although functioning very much as a free (at point of use) in-house scientific consultancy, the world was changing, and the prospect of privatisation cast the work of Scientific Services in a financial spotlight. It became a self-accounting unit, changing its name to the catchier Scientifics, with internal charging for its services. In the real world a depot engineer was pleased to find out, by a telephone conversation with the lab, what the most durable paint for a pit floor contaminated with a specific lubricating oil was. But when he had to pay for that information, it suddenly seemed less important. By the time it was sold in 1996, Scientifics had its own accounts, personnel and marketing departments, and sales and contracts managers. Chemistry was what brought in the money but not necessarily what solved the problem. When offered for sale, Scientifics was cast adrift from the rest of BR Research and sold as a stand-alone entity, joining the world of the private scientific consultancy community. It is to its credit that, following privatisation, staff successfully demonstrated their skills to a wider market. But in so doing, the idea of the dedicated railway chemist — part of the railway family — had gone. The depot manager will, metaphorically, look up Yellow Pages and not the internal BR telephone directory. The solution to his problem may be offered by someone who was dealing with aeroplanes yesterday and cars tomorrow, not someone who had been dealing with trains last week and would be next week.

But our history lives on.

CHAPTER 2

The Area Laboratory

*John Sheldon, with help from a few friends,
introduces us to the Area Lab, some of the people
who populated it and the work undertaken.*

THE BEDROCK of Scientific Services was the Area Laboratory, where the routine work of the chemist assisting in the smooth running of the railway was carried out. Using the Derby Area Laboratory as the main example, follow me and some of my colleagues who entered the laboratory for the first time more than 60 years ago, as we get to grips with the work which eventually allowed us to become proficient Railway Stinks. But first, a stirring story of what could be involved in the humble trade of the railway chemist.

Pictures 1 to 8 in the photo section of this book shows some of the work carried out on a daily basis by Glasgow Area Laboratory scientists. Photos 9 to 17 relate to Derby and Doncaster laboratories.

The Cold War against communist Russia had been raging for more than a decade when two railway chemists from the Derby Area Lab were ordered to Trent Junction, a key railway location to the south of Nottingham. Ordinary lads, not heroes, they had been briefed in advance and knew what they had to do. The sheet and sack stores building, a former Anderson air raid shelter, had been infiltrated and their job was to ensure that there would be no living thing remaining by the time they left. Their weapon of choice was trichloroethylene, known in the trade as Trike, a toxic solvent. The pair were young and keen, knew the dangers and planned for them. One would enter the building by its only door, fully kitted in overalls and a gas mask, and make for the far end. Then he would open the drum of Trike and liberally disperse the contents over the opposition as he made his way back to the entrance. The other operative, similarly kitted up, would stay at the door, ready to intervene at the first sign of his colleague being overwhelmed.

In the event, the opposition kept a low profile. Operative One's only hindrance was having to climb over the empty grain sacks, piled within three feet of the roof, in which the enemy had secreted themselves, without disturbing them. Reaching fresh air, the chemists slammed shut the door of the store, cutting off any escape, removed their protective clothing and relaxed for the first time since arriving on site. Neither knew, would never know, how successful the operation was but on return to HQ, they reported having dispatched hundreds, if not thousands, of the evil grain weevil. The West was saved: once the solvent evaporated, the sacks could be used again, and vital food supplies would continue to flow by rail. Further research showed that 1.1.1. Trichloroethane (aka Genklene) was as effective and presented less risk to the operatives but modern health and safety regulations would rule out the whole process. The current advice is that trichloroethylene has a short-term exposure limit of 150 parts per million, that it is carcinogenic and can cause skin irritation. Their parting words summed up these brave lads: "Same time next month. I'm off to the pub."

I was one of those who undertook the disinfestation process at Trent Junction. While subsequent chapters will explore the more specialised areas of the work of Scientific Services, the typical route into specialisation was via a thorough grounding in the Area Laboratory. Scientific Services has always been populated with 'interesting' characters, so please indulge me in thumbnail sketches of the colleagues I came across in the 1950s and 1960s. Eccentrics and their eccentricities did not die out in the 1960s, of course, but to save embarrassment we shall not dwell on those of later generations!

The term 'area laboratory' was coined since laboratories were located throughout the UK in recognised railway towns. The railway industry has always been a vast procurer of various products which must meet specified requirements to ensure they are fit for purpose. The area labs were initially created to make sure the products were to specification and suitable use by for chemical and/or physical testing. Indeed, any problem that could be resolved scientifically usually ended up at their door. Routine materials subjected to analysis by wet chemical techniques or physical testing on a regular basis included steel, brass, white metal, paint, water, cleaning materials, coal and lubricants, to mention but a few.

On entering the Calvert Street edifice on my first day, I was greeted by Tommy and introduced to Jimmy Harold, an avuncular Scot and chief clerk, who would serve me well in years to come by ensuring that my part-time education in appropriate chemistry qualifications was uninterrupted by National Service. I also met the area chemist, Stanley Bairstow, a small balding Yorkshireman exuding boundless energy. I couldn't help but notice two fingers of his right hand were webbed. Mr Tomlinson — addressed formally when dealing with him directly but known as Tommy among staff — issued me with a pristine white lab coat, the standard uniform of the bench chemist, and directed me to the oil bench in the main laboratory, my first workstation under the tutelage of Thelma Allison, a petite twenty-something. The oil bench was hardly my idea of chemistry, since testing oils involved physical properties: viscosity, specific gravity, flash point and pour point. But it was a start and I was a proud working lad. Many years later, after the closure of Calvert Street and the subsequent move to Hartley House in November 1964, I was put in charge of the oil bench, which was still the introductory route to Scientific Services work for many a new recruit. History repeating itself!

From the oil bench initiation, I was gradually introduced to other, more chemistry-oriented work involving the analysis of paints, steels, non-ferrous metals, waters (drinking and boiler). To ensure that all the area labs were singing from the same hymn book, as it were, a mighty three-volume tome referred to as the 'bible' was issued to new entrants, which documented analytical methodology for everything the railway chemist was likely to encounter.

George Haynes was the firebrick tester. In these pre-diesel and non-electrified days, the railway was entirely reliant on steam locomotion and testing the performance of the firebricks to be used within the firebox was essential.

The steam locomotives were fitted with a brick arch between the firebox and tubeplate to improve combustion. Tests on the bricks were carried out to measure dimensional change and stability. Two-inch cubes were cut from the bricks using a water-cooled carborundum disc.

Cubes were heated to about 1300°C in a gas furnace, allowed to cool and measured to quantify any unacceptable expansion, contraction or other defects. A small cone cut from the brick was heated in an electric furnace to determine the melting/fusion point. These tests helped to prevent premature failure of the arch which could disable the locomotive.

Ken Phillips and Alan Astles were jobbing chemists. Ken left about a year later to become a lecturer at the local technical college, while many years later, Alan became Area Scientist at London and sadly died of a heart attack while in post. John Hudson, a Heanor lad with an accent to match, was with us for many years before taking a job with a borough council in Yorkshire as a sewage specialist. Subsequently meeting up with Alan and having been made aware of Alan's promotion to the London lab, he memorably remarked: "Ay-up Alan, so you'll be in the smoke and I'll be in the shit!"

The so-called paint girls arrived from former accommodation in Cavendish House to be integrated into the Calvert Street regime about a year after my arrival. Pat was charge hand over Wendy, Beryl and Audrey. They brought with them their boss, a large gentleman called 'Nobby' Hall, which was just as well for I don't think Tommy would have been able to cope with a sudden influx of feisty young ladies. Cavendish is the family name of the Dukes of Devonshire, and Cavendish House originally existed as overnight accommodation for the family before boarding the Midland Railway train service to London St. Pancras. As a part of BR Research in addition to the paint section, Cavendish House also housed the Physics and Textiles Division, the latter responsible for testing upholstery, carpets, uniform materials, wagon sheets and so on.

Back at the lab was Tom Brown, who as a child contracted polio and hence walked with a pronounced limp, and water analysist John Gessey, who worked in the appropriately named wet room. Phil Dimmick had returned after two years' National Service. Phil was a member of the Plymouth Brethren Church, of which I had no knowledge. He and his wife, although having a young family of their own, became committed foster parents for many years.

Steels and non-ferrous metal analysis took place in a lab separate from the main one, which was occupied by Fred Eccleshall, an elderly bluff gentleman rich in corny jokes, Anne Wilkinson and John Fisher. Anne was the lab sweetheart with whom I had a snog after a Christmas dinner. My romantic aspirations were dashed when it transpired that she had a regular boyfriend in the RAF. John was an odd character who led a somewhat vagabond life and smoked dog ends. His wife lived in Guernsey while John lodged in a B&B she owned in Matlock Bath. He visited his wife on a monthly basis and slept rough en route. Tom Randal was a sandy-haired fellow with a certain military bearing enhanced by his moustache. To me, his appearance over the years never seemed to alter and he made a career of seeming to do not very much.

Bob Meikle, the most senior member of the lab staff, was a Scot who returned to his roots by virtue of promotion to the post of area chemist at Glasgow. Ossie Frearson, the water and environment specialist, always had his fishing rod to hand when taking samples from water courses and ponds adjacent to rail tracks and depots. Jim Ward, a year older than me, had been in post for a year and Tommy instructed Jim to take me to Derby Tech to sign up for advanced levels that ordinarily I would have done in sixth form, thence paving the way for a part-time degree. However, the road to hell is paved with good intentions and my further education aspirations didn't quite go as planned, so I had to settle for the National Certificate qualifications.

Way back then, laboratories used a vast amount of glassware — beakers, flasks, test tubes, retorts and more — and all of it required cleaning. Dishwashers were not an option in the 1950s, so a lady by the name of Maggie Mather was employed for this purpose and handwashed each item.

Separated from the main building was the so-called rough lab, which served as storeroom and a place where chemical solutions in bulk were manufactured for use in Derby's Locomotive and Carriage & Wagon Works and outstations. Bill Sneath was its guardian, as well as landlord of the Brunswick public house nearby.

Occasionally staff were required to leave the laboratory confines to attend to jobs of a routine nature or ad hoc problems considered to have a scientific solution. As well as killing grain weevil, such jobs included oil sampling. This is when routine oil samples for lab testing would be collected at depots and outstations and brought to the lab by Alf 'Robbie' Robinson. It was his full-time job; his knowledge of the BR timetable and local bus services rivalled that of the London cabby. When he retired, he left his book of knowledge to those who came after. This job fell to a junior but was short lived and considered inefficient, so it was left to the depots to dispatch samples direct to the lab via the BR internal parcel post system.

The reclamation lab, or rec lab as it was colloquially known, was a satellite manned by two unsupervised staff members considered competent and mature enough to perform the duties demanded by this posting, which was never no longer than two years' duration. The building was a two-storey brick structure in the heart of the Carriage & Wagon Works and attached to Q Shop, a foundry that reclaimed brass and white metal (alloys containing specific metals to improve their performance) from worn wheel bearings of rolling stock. Castings produced from the melts were then analysed. The senior of the two chemists was responsible for brass while the other dealt with white metal. Sample ingots arriving from the shop next door were prepared and analysed by traditional wet techniques. Results were logged, reported by phone to Tommy and to the works metallurgist who would calculate the quantities of constituent metal elements to be added to the melts to produce the right properties for the bearing alloy being produced. Being unsupervised, Tommy would pay an occasional surprise visit to ensure we were running a tight ship. Generally, we were never surprised, since we had prior knowledge of visits from a friend back at base.

My first rec lab posting was with Phil Dimmick and once I was trained up to the arcane science of white metal analysis, my year went without incident. Phil, about five years older than me, was an amenable chap and we got on well as a team; a necessity in our cloistered circumstances.

My second posting was with Andrew Mailer, whose main claim to fame was as the son of the Chief Medical Officer for the Scottish region. A second, unsubstantiated claim to fame was that he was propositioned by the TV personality Gilbert Harding on the night sleeper out of Glasgow. Despite being of a deeply religious persuasion, he was nonetheless tolerable company for an atheist and again a bit older than me, so got the brass analysis. Of the four constituents of brass, the minor element was calculated by difference, having ascertained the other three analytically. Andy, in the guise of efficiency or expediency, calculated both minor elements by difference, leaving just the copper and zinc to be determined by bench work thus eliminating the tedious and time-consuming business of performing three analyses. Expediently, this would give him more time to spend on motorcycle maintenance.

I suggested that rec lab staff were chosen not only for their competence but also their maturity. What I am about to reveal may question the latter but bear in mind that we were still relatively young. Since the daily routine required the analysis of one melt per day — i.e. one brass and one white metal — we were left with time on our hands and as the saying goes, idle hands make mischief. Works metallurgist Jim Jackson had a young assistant called Mick, with whom Andy and I had good rapport. Nonetheless, we didn't let that get in the way of practical joking. Access to his upper floor office was gained by a fire escape stairway and Mick was regularly ambushed mid-flight by water discharged from a redundant fire extinguisher. Mick exacted his revenge in spades, or should I say buckets, as he emptied the contents of a two-gallon bucket of water between a crack in the wall/ceiling interface, most of it saturating our shared desk and paperwork. End of water wars!

We also had a good relationship with the lads in Q Shop but once we discovered how to manufacture nitrogen triiodide, a compound which explodes on contact, we placed the substance on the heavy wrought iron latch that secured the shop door when it was closed for the day. We ensured, by a dummy run, that the desired effect — a significant bang as the latch was lifted — was achieved. On entering the lab the next day, we discovered a broken pane of glass and a 10 kilogram chill (a water cooled mould used for rapidly cooling a casting) on the lab floor. Nothing was said by either party but there were no more jokes.

My last partner was Peter Millband, coincidentally an old boy from my school. He was a bright and engaging young fellow with a tendency to the avant-garde. By now the senior partner, I was the brass man and Peter white metal. While perfectly competent as a chemist, his heart and mind were not really suited to the subject. Pete had a penchant for mixing chemical compound solutions on paper, drying them and hanging them on the wall, Jackson Pollock style. He moved on to pursue a career in the arts and subsequently achieved fame as Young Artist of the Year in 1964.

Meanwhile, back at Calvert Street our minds were occupied by waste disposal. The 1950s was a time well before the advent of the Health & Safety at Work etc. Act, COSHH (Control Of Substances Hazardous to Health) and the recycling ethic, and as such, waste disposal was very much an ad hoc operation… a case of 'needs must when the devil drives'! Waste flammable solvents such as xylene, toluene and acetone were collected in bulk and periodically burnt on a dedicated square metre in the quadrangle that separated the rough lab from the main building. The only nod to health and safety was to close adjacent windows.

On one occasion there was a certain amount of alarm and despondency when a chlorinated solvent contaminated the flammables during a firing, creating phosgene, infamously utilised in the First World War. Prompt application of gallons of water quickly resolved the problem, avoiding the need for a mass evacuation of the immediate neighbourhood. To the rear of the lab was the termination of a disused arm of the Derby Canal which, when operational, serviced the former power station. Now, as far as the area lab was concerned, it was a legitimate dump for redundant chemicals, no matter how toxic. The highlight of this method was disposing of sticks of sodium metal, which would emit a satisfying explosion accompanied by a waterspout. In hindsight, I often wondered about our potential use of sodium metal.

The area lab also manufactured chemicals for use on the railway. The Loco Works power station required large volumes of the chelating agent EDTA (ethylenediaminetetraacetic acid) for water softening. This was prepared in the rough lab by dissolving the specified amount of the solid powder compound in 25 litres of water in glass carboys. Clive Hickman, a somewhat accident-prone young man, was nominated for the job. At this time 50 litres, or two carboys, were required and vigorous agitation of the carboy was necessary to effect solution. In the promotion of efficiency, our intrepid Stink decided to agitate both carboys at the same time, rotating each in either hand. Naturally the inevitable happened.

The carboys made violent contact, resulting in a flooded rough lab. Needless to say, Clive was never invited to take part in this activity again. Filter papers, integral to many lab procedures, are not always the ideal filtration medium and often the use of filter paper pulp is preferred. This is manufactured in the lab by immersing filter paper scraps in boiling water on a hotplate in a fume cupboard until a pulp consistency is obtained. Clive, while making the pulp, decided it was a good idea to hasten the process by securing the neck of the flask with a rubber bung. As the pressure within the flask increased, the inevitable happened. Fortunately, the integrity of the flask was maintained in favour of the bung, but the fume cupboard had a pebble-dashed appearance. Alan Astles pertinently remarked that Clive was merely testing the veracity of Boyle's Law.

Fog signals, also known as detonators, were routinely tested. For the benefit of the uninitiated, fog signals were placed on the rail to warn the driver of a hazard ahead or a red (stop) signal, which in foggy conditions cannot be seen in time to take appropriate action. To all intents and purposes, they are mini bombs emitting a loud bang as the train passed over them. They consisted of a quantity of gunpowder and detonators (percussion caps) in a round metallic container about the size of a jam jar lid, with lead strips to attach them to the rail. The top and bottom of the container were crimped together to make a hopefully watertight seal.

At the lab, a sample would be deconstructed by carefully opening the container: gunpowder would be weighed to ensure the specified quantity was present and the percussion caps tested for functionality. They were also immersed in initially near-boiling water for 24 hours to check for waterproofness. Finally, a random ten would be site-tested under working conditions. The chosen site was a small length of track in the loco works adjacent to the typing pool, and a wagon would be winched over the test samples.

Prior to this exercise, all interested parties had to be informed, and this included the ladies of the typing pool — after all, a series of loud explosions close by could be upsetting. Informing them of an upcoming disturbance could be daunting for a shy lad on the receiving end of some rather unladylike comments. Anecdotally, a farmer attempted to obtain compensation from British Railways by alleging that his sow in fallow died from shock due to fog signals being detonated on the line adjacent to his land. He was not successful! On a more sombre note, a track worker was seriously injured by a fragment from a nearby detonation. This unfortunate incident prompted a change from metal casing to one made of plastic, which proved to be an unsuccessful experiment. The solution was a redesign of the crimp to reduce risk of fragmentation.

Bob Symcox, then at the Crewe laboratory, comments that on one occasion, a newly employed member of staff was getting a bit bored by the required restrictions when deconstructing a fog signal. Using a file to open the container was a slow and laborious process, and the lad thought it would be far quicker to use a grinding machine. Luckily, he was restrained before the resulting shower of sparks reached the gunpowder! Vince Morris, recalling his days at the London laboratory, tells of a novel use proposed by the driver of a diesel locomotive. Each cab was equipped with a supply of fog signals securely stored in a metal tube. This chap had taken a dislike to a fellow driver and, knowing that this gentleman would take over driving the locomotive, he purposely raised the position of the driving seat. He achieved this by winding up the scissor mechanism beneath the seat, ensuring that his intended victim would have to wind it down again to restore a comfortable driving position. He inserted a detonator within the scissor mechanism so that as it was wound down it would be squeezed — and explode. Luckily it didn't, since there was insufficient percussive force.

Samples of creosote, used to preserve wooden railway sleepers, arrived on an almost daily basis to check it was up to specification. It was probably the least liked of all routine testing. Nasty, smelly stuff, creosote's residual odour would cling on. Arriving home, my dad would say: "On creosotes again!" If one of the Stinks incurred Tommy's displeasure, about half an hour before home time he would deliver a crate of six cans and insist they could not wait until the following day, despite them having sat in his office for a couple of hours. All that was required was the water content and specific gravity, so it turned into a game of 'beat the clock'.

In those early days we worked a 40-hour week, subsequently reduced to 38, so we stayed behind on Monday evenings to make up the hours. This was without any senior supervision as Tommy always left at five o'clock sharp. Left to our own devices, the firebrick lab became a sort of youth club. I would sometimes bring my guitar and singalongs would occur. Occasionally these became rather noisy affairs, and the sound would carry through the main lab to the office from where Jimmy Harold would good-naturedly ask us to tone down the volume. Card schools flourished and another pastime evolved: guess the volume. This involved filling assorted vessels of varying capacity with water and asking everyone present to estimate the total volume. The most accurate guesstimate collected the kitty.

My memories of starting at Calvert Street are reinforced by Dave Smith and a contemporary of his at Derby, Roger Hughes, who later became the Area Scientist at Glasgow. Dave now takes up the story…

My first day at the lab was in July 1962 and I soon met many memorable personalities, but my first thoughts were 'you must be crazy to think of working here'. When I was ushered towards my bench, I noticed that virtually no one had a hole-free lab coat and I was left wondering if this was standard issue or if the moths were exceedingly large. Unlike John's experience when he started, the coat I was given was full of holes and many stains because it was a cast-off from someone who had left (or had possibly been poisoned?!).

A few minutes in the lab with whirring extraction fan motors (the fan blades had long dissolved with acid fumes) and watching my new colleagues at work, I realised I had entered a very strange world! On being shown round the laboratory, I noted that before I shook the hands of those welcoming me, they removed pieces of rubber from their thumb and index finger, often to reveal brown skin from exposure to nitric acid or other chemicals. These pieces of rubber came from hoses and were slit to accommodate the digits and protect them from hot flasks. By day two, I also had brown fingers. The women probably suffered most from the atmosphere because their stockings gradually dissolved over the day as acid or solvent spots came into contact with their legs. The place had its own characteristic smell, but your nose became conditioned to it and you did not notice after a few days. It was obviously unhealthy because germs did not breed: you never caught a cold in this working atmosphere.

Many of the facilities were long past their best, as were some of the occupants who were part of the old school. In its heyday, the railway chemist had significant stature and our labs had evolved from those early days. This was the place John (the author of this chapter), John Hudson, Brian Leeson, Jim Ward and their like inhabited, together with the legendary Heinz Bauer, an Austrian Jew who came to the UK to escape the fate that befell many of his friends and family. John recalls that he first met Heinz in his role as lab attendant at the Derby Art School where, somewhat bizarrely, John went for his day-release physics classes. Heinz had a dramatic impact on those around him with his philosophy on life, liberal views on sex and a very different attitude to food. Continental sausages, such as cabanos, were his mainstay and yet in those days very few of us had ever even seen them.

The laboratories typically looked like the ones in films, where mad chemists (Dr Hyde comes to mind!) make potions. Perhaps some of those who designed the film sets had worked in our labs? Calvert Street had wooden benches with shelves full of chemicals, and there were beakers and glass retort vessels everywhere. The room I was in was mainly used for the routine analysis of metals, and although not quite as smelly as the main lab, with its many heavy organic vapours, acid fumes were dominant. Depending on what we were analysing, all the nasty acids were used alone or in some deadly combination virtually every day.

It was in this room that I sat opposite Heinz Bauer and worked with Alan Baker, Clive and the two Toms — Tom Randall and Tom Brown. They showed me that some of what I thought was essential laboratory equipment had more than one purpose. Ascorbic acid was an essential reagent for some occasional analytical processes, but it was taken daily by the teaspoon as a preventive measure against illness by some — vitamin C by another name.

The drying ovens were used for lunch, so pies and tins of beans were cooking, and chips were kept warm alongside chemicals that needed to be dried and weighed. Heinz used the ovens to heat up cabanos stew in a Kilner jar. An illustration of the lunacy adopted is the case of Clive and his tin of beans. The tin was put into the oven, unopened for a good reason, but no one could be close by when he opened it as a hot stream of bean juice would shoot out over some distance. The beakers we used for analysis doubled as teacups and my spatula served the dual purpose of mixing chemicals and stirring tea. Only Heinz had a mug he treasured, so much that one of his colleagues stuck it to the middle of his bench with epoxy adhesive — and there it had to remain!

The work carried out very much reflected what the railways did in its heyday. The old BR even had hairdressers as well as hotels. The railway works, where they built locomotives, coaches and wagons, also made their own cleaning materials equivalent to the old scouring powders, paints and greases. Whenever there was a problem, the works staff turned to the chemist — sometimes pronounced with the 'chh' sound. People would turn up on our doorstep with a sample for us to sort because of some undefined, or at least under-defined, problem. Unfortunately, the samples were not always at their best. A sample of axle box grease in a brown paper envelope was not at all helpful if they wanted us to determine why it was not working, since most of the liquid was absorbed in the envelope. Providing a bottle of tap water which 'tasted funny' in a salad cream bottle still containing visible amounts of salad cream was also a challenge. Additionally, an immediate answer was usually expected. Often, in such cases, we had to visit the location to collect the samples ourselves in clean containers. We analysed pork pies, paints, oils, wine, coal, cleaners and whatever was a problem or was deemed necessary to be controlled.

In my laboratory, Tom Randall and Heinz Bauer did such interesting work and investigated claims made against BR for damaged goods. The more advanced work was left to those in the adjoining 'red room' of scientists who were at management level. Some of their analytical processing seemed to be based on the Sherlock Homes approach — pipe in hand and plenty of pontification but not necessarily much dynamic action. This contrasted very much with us lot slaving away, trying to analyse steels and brasses within 24 hours of receipt.

"Which beaker has my soup in it?"

Apart from putting on water baths, the imaginary fans in the fume cupboards and making coffee and tea, there was another morning ritual that had to be followed. This event was Tommy delivering the platinum which was stored overnight in a giant safe. In all probability, the value of the platinum exceeded the value of the building. I used to receive a lot of it in the form of electrodes and crucibles. We all were issued with a piece of platinum wire as a personal item and in my case, it was in sets of two round cylinders of meshed platinum, one being smaller. This meant I was able to attach it to ancient electrical equipment and use them to electroplate the elements copper and lead onto the surfaces. The analysis of brasses required me to weigh the sample and dissolve it, and then deposit the metals on the pre-weighed electrodes to determine how much of each metal was there. Tin was determined by precipitation and the total always came to 100 per cent, as we determined the zinc content by difference. This was indeed fortunate because Tom Brown was responsible for the analysis of white metals used on axle boxes. His analytical process required him to add a small amount of zinc to help the zinc precipitation process to proceed. He then subtracted the one per cent zinc he added from the total amount he weighed in the precipitate. Tom, being an honest person, often reported a negative content of zinc as he found less than he started with!

Tom Brown had reputedly contracted polio in his youth and therefore dragged his foot slightly. He worked in a little offshoot to our lab, with a chimney as his extraction system. He used hydrobromic acid, which produced thick dense brown fumes more pungent than chlorine, its close relation. When disaster struck, you would hear Tom coming, dragging his foot and shouting. We all knew then it was time to evacuate the building fast, as Tom appeared with thick brown choking fumes a few feet behind him. This was the only time that the other Tom, Tom Randall, moved, it would seem. He sat, as we all did, on high-backed stools, except he rarely seemed to move unless he heard Tom Brown coming towards us. He was even known to leave his cup of tea behind.

As my time progressed, I had greater opportunities to see the full picture which made my induction very worthwhile. Many years later I had time to reflect on this startling beginning. Life at Calvert Street was a good training centre — providing you survived the apprenticeship. It was to prove the foundation stones for dealing with the 'wrong sort of leaves', the 'wrong sort of snow', the 'wrong sort of toilets' and the 'wrong sort of fuel' to mention just a few noteworthy adventures included in later chapters.

Roger Hughes began working at Calvert Street a few days after Dave Smith. After he was 'kicked out' of Nottingham University, he went to the Derby Labour Exchange and made the rather unlikely claim that he was looking for a job on a whaling ship. However, the clerk was helpful and gave him the address of the Seaman's Union in Liverpool, which eventually led him to be accepted by a whaling company. But as their fleet only sailed once a year and had just left, there would be a delay before he could start work. So, he went back to the Labour Exchange, now declaring himself to be an unemployed whaler and therefore entitled to dole. He met the same helpful clerk, who told him he would have to apply for jobs and provided details of several vacancies. After interviews, Roger got offers from four firms but chose the railway while he waited for the boats to come home. He turned up at Calvert Street and met a relatively new starter whom he recognized. It was the clerk from the Labour Exchange — none other than Dave Smith, who knew a good job when he saw one!

While Roger's early career followed the pattern of John and Dave, here he writes about the general description of the work at Derby in the 1960s.

Tommy (L G Tomlinson) told me I would start on the oil bench. Jimmy Harold, the chief clerk, gave me two lab coats and told me my bell code. I think it was three longs and one short. Tommy used a bell to summon lab staff to the office for instruction or reprimand, and everyone had a different code. The lab coats soon had two corporal's stripes in ballpoint pen ink on the sleeve, holes burned in them and oil stains down the front. The rear seam was stapled together from the top, because when going upstairs two people had grabbed a tail each and ripped the seam to the neck.

The senior technical assistant in charge of the oil bench and bearing a startling resemblance to Adolph Hitler (particularly the moustache) was Aubrey Dunne. Aubrey explained my duties, but from the beginning I was a bit put off by his habit of pretending to spit on a small area of the workbench. He never actually spat, and the reason only became clear on the Friday afternoon. The last hour on Friday afternoon was clean-up time, when all the apparatus, work surfaces and so on were cleaned, and reagent bottles dusted and filled up. The only area left untouched was these few, detested square inches on the workbench. It transpired that Dr Beeching had leaned on this part during a visit.

It was the weekly job of the junior staff to make up and refill the bench reagent bottles. This involved, without any protective equipment, manhandling and pouring concentrated acids and alkalies from Winchesters and carboys into smaller containers, then diluting the concentrate to bench reagent strength. No wonder our trouser legs had lots of small holes in them. Another weekly job was to get a porter's barrow and trundle a drum containing several gallons of waste flammable solvent down the street, across the main road, under the main rail line and onto waste ground in front of the workshops. The solvent was poured into an open 45-gallon drum and ignited. You started off about six foot away, lobbing lighted matches towards the drum. As each attempt failed, you moved closer and closer until you were holding a lighted match just over the rim. There would be a very warm whoosh, and you would jump back minus your eyebrows and fringe.

Arnie Wooller was a phenomenal qualitative analytical chemist who had worked all over the world. He relied on burning and smelling a sample, then, following a small card index box with analytical methods scrawled in minuscule writing, he would, with a couple of test tubes and few basic reagents, pronounce the composition of a mixture. In the upstairs attic lab, he taught me how to make the explosive 'gun-cotton'. During the lesson, there was a minor explosion in the oven which caused the door to buckle outward. I had the door on the floor and was jumping on it to get it straight when Tommy walked in. To his credit, he just tutted, turned around and went back down the stairs — and my bell never rang.

In the early 1960s we had very little sophisticated equipment. Routine calculations were done in longhand, checked the same way and signed off by a colleague. I still have lab notebooks from that era. Eventually the oil bench was allowed to use a three-foot slide rule to check calculations. Latterly the lab got a Hewlett Packard calculator that could add, subtract, multiply and divide. This was kept in a safe and was signed out each time it was used. One day Tommy proudly produced an oscillating electric hotplate, but this did not survive long, after Clive Hickman started using it to heat his stew. The inevitable happened: the container broke, and the stew penetrated and destroyed the delicate electrics.

All analysis and testing required glassware, but this was before custom-made systems with ground glass joints. Glass blowing was a necessary skill. When I first joined, I remember being shown a beautiful glass swan made by Stan Bairstow, who went on to be the head of Scientific

Services. You filled it with water, blew down the tail and a jet came out of the beak. I tried and got soaked because there was a small hole at back of the head, which the knowledgeable covered with their thumb when they blew down the tail. We had glass tubing and cork and rubber bungs to manufacture connectors, joints and seals. We used to cut the tubing to length with a broken grey ceramic heating electrode from one of the furnaces, smooth the ends using a Bunsen gauze-mat like a file, then heat the ends in a flame until they began to melt. When cool, these ends would be smooth. We then had to bore a hole in the cork or bung and ease the tight-fitting glass tube through. Accidents inevitably happened. Once a long piece of tubing broke as I was pushing it through a rubber bung, and it went into the palm of my hand and out through the back. Fortunately, it came out easily and with no lasting damage. The joints that resulted from the use of corks and so on did not always have great integrity, and one piece that suffered was on the end of the oil bench: the five litre still used to recover flammable solvents used to clean the apparatus. It caught fire with great regularity and as a great excuse to use the carbon dioxide fire extinguisher.

When we first moved to London Road, I was so annoyed that some miserable manager had installed frosted glass in the bottom two feet of the window that I spent ages putting clear tape on it, so I could see out towards the RTC opposite. Even though we were young, many of us were brought up in the old school of wet chemistry and were suspicious of the heavily instrumented analytical group downstairs. We had spent years perfecting laborious and meticulous analytical techniques. Hours of effort resulted in an analysis, reported to perhaps two decimal per cent places, with maybe up to '10% SiO_2 by difference'. This was replaced by Heinz squirting a few X-rays at the sample and giving the analysis in parts per million in seconds. Fortunately, we got on well with the staff and one job in particular discharged any lingering resentment — wine testing. With its hotels, ships, station restaurants, dining cars and so on, British Rail was a major wine seller and held vast stocks, stored underground at major stations. The wine tasters were looking for chemical indicators to relate to their tasting notes and to assist in predicting changes in the wine during storage. The newly available analytical technology at Derby seemed to be the answer, so dozens of bottles arrived for analysis. Sadly, these new-fangled instruments only needed a few micro-litres for the test, leaving 99.999 per cent of the bottle's contents as waste to be disposed of. The Analytical Unit staff were pleased to accept the help of the upstairs staff in the disposing of this burden, thus cementing a strong relationship between the groups.

Scientifically I can't remember many major cock-ups but I do recall one. Dave Panett arrived from British Steel as the 'the man' for steel analysis. All went well until errors were found in reported analyses. It transpired that Dave was using conversion factors from his steel manufacturer's pocket diary, which contained errors. What an embarrassment!

Before health and safety legislation, the railway was an interesting place to work. The only safety instruction we had was "be careful." The only protective clothing was our cotton lab coats and 'fingers' made from a couple of inches of rubber tubing slipped over our fingers, so we could lift beakers of boiling acid. You could always tell the steel bench staff by the tips of their fingers, burnt shiny and brown by the nitric acid. The health support for those in the workshops was of the same quality. I remember giving an asbestos lecture and the front row all had cupped hands to one ear. They were boiler makers and smithy workers suffering from noise-induced hearing loss.

One of our jobs was to test the Perliton carburiser used in the workshops for hardening steel components. Items were dipped into a bath about three feet square containing molten sodium cyanide. This was done in a corrugated iron shed with no forced extraction or ventilation. The workers had blue overalls as protection. Once I asked one why he had a large plaster on his head and about ten square inches of hair missing. He explained how he'd put a component that must have been damp into the vat. The subsequent explosion resulted in a lump of molten cyanide burning his head. He visited the works nurse, who gave him Germolene and a plaster. I followed this up with the nurse, who said he never mentioned cyanide, before nor after.

I had another run-in with the nurse when working alone through a lunch hour. I was making up a test tube of five per cent potassium cyanide solution reagent. For safety reasons I decided not to use the usual technique of putting your thumb over the end of the tube while shaking it, using a rubber bung instead. Unfortunately, I squeezed the bung in too hard and shattered the tube, which cut the palm of my hand. We had a saying that if you can count to 20 after cyanide poisoning, you will survive. Counting furiously, I grabbed the antidote bottles, poured the contents into a beaker and swallowed the lot. Those of you who have seen the result of mixing the antidotes will wonder how anyone could drink a mixture that looks like blue-green snot. Fear can make you very brave. After the incident I phoned the works nurse, who said there was nothing she could do. "If it was serious, you'd be dead," she said.

Despite all the hazards, the one thing we were most worried about was telling Tommy we had lost, broken or destroyed a piece of equipment. Two incidents come to mind. Firstly, analyzing an 'unknown' and taking the bottom out of a platinum crucible, because the unknown contained lead, and secondly, when the development of anti-icing grease for switches led me to working with the Derby South track gang, testing grease in-situ and measuring the switch plate temperatures in the middle of winter. The entire gang was Polish, made up of men who escaped German invaders and came to Britain to continue the fight. Following the war, they could not safely go home. Heavy snow was falling, and I was concentrating on measuring temperatures on one of the tracks south of the station. I heard a horn, looked to my left and saw a train a couple of tracks up from me, then went back to measuring. The next second I was yanked by the collar as a loco ran over the temperature probe. Johnny Beibek, the gang foreman, had pulled me from under the wheels of a loco going into the workshops. I bought the whole gang a pint in The Vic before going to face Tommy with a bent and flattened probe.

Stories from Crewe, Doncaster and London show that the experiences of Roger Hughes in Derby were not unique in the annals of Scientific Services. Bob Symcox, who finished his career at the Swindon lab, recalls a couple of incidents at Crewe and Doncaster.

We moved out of the old Crewe laboratory in the centre of town in 1968 to establish ourselves on the twelfth and top floor of Rail House, overlooking the station and associated rail tracks. Before we left the old place, we cleared out chemicals and kit that were old or no longer needed.

Pat Edwards had been working on a problem with the gas-fired LPG point heaters installed on the mainline junctions in the Crewe area. It was determined that there was corrosion of silver contacts taking place in the ignition system, which was preventing the LPG gas being lit. This was allowing snow and/or frost to build up on the points, rendering them unusable and therefore delaying trains. Pat obtained a sample of the stenching agent (ethyl mercaptan) used to make LPG smell and confirmed that this agent was indeed the cause of the corrosion, and the problems were rectified. However, a bottle of the agent — I think 500ml in volume — was found in a cupboard one morning and someone decided to pour the contents down the drain. Soon after lunch, the local Gas Board were seen diverting traffic in the road below and digging holes in several places. The penny dropped! Someone, I forget who, took the brave step of going downstairs to speak with the workmen, who certified that the smell was from an open drain gully in our yard and not from their pipes.

Early December 1981 brought very cold weather, with snow in the first week, severe overnight frosts and temperatures barely reaching zero centigrade in daylight. This was followed by a very wet period, causing flooding in parts of Yorkshire and so many areas were frozen solid. On 8th December, a York-to-Liverpool train derailed as it crossed the bridge over the River Wharfe on the approach to Ulleskelf Station. Once the train was recovered and the track reinstated, I was sent to the site to set up a temperature recording system at the request of the Eastern Region headquarters in York. A van with three members of the local permanent way gang collected me from the Doncaster laboratory. Once we left the main roads, conditions became increasingly treacherous due to frozen snow and ice, together with melting ice where road salt was beginning to take effect.

About a mile from Ulleskelf, we came upon a car stuck in the middle of a country lane with the driver still at the wheel. We got out of the van and, pulling on our Wellies, approached the driver, who was revving the engine and getting nowhere! Broken ice from the road had become packed under the engine block, which had forced the road wheels off the ground — so it was no wonder that the poor man was stuck. A bit of brute force soon got the car and motorist back onto the road surface, and we cleared enough ice from the road for him to continue his journey. Back in the van, we followed him to Ulleskelf, when he went to the old goods shed where he had a business. We went about our task, setting up the temperature recording equipment on the track, and returned to our vehicle. When the permanent way men phoned in to say we were done, they received a message that they were needed elsewhere and so we agreed that they would take me to York Station so I could return to Doncaster.

Just as we were leaving, the motorist we had rescued came running out of the goods shed and presented each of us with a large bouquet of flowers for rescuing him. As he was a wholesale florist from Holland, he said it was the least he could do! You could just imagine the looks I got at York and on the train, dressed in track gear with Wellington boots and rucksack and a large bunch of flowers. In retrospect, that was the easy part of the day. When I got home, my wife took a hell of a lot of convincing as to how I came by the flowers. But as it is said — that is life.

Ian Cotter began his career at the London laboratory in the 1960s and worked closely with BR shipping services, a legacy of the pre-nationalisation railway companies' fleet of steamers and augmented by the more modern roll-on, roll-off vehicle ferries. The ferries were mainly diesel powered as the steam turbine ships were phased out. However, diesel fuel was more expensive than the heavy fuel usually used in boilers; mainly the thicker residue from crude oil distillation. Thus, many marine users went over to using boiler fuel in their ship's diesel engines. To enable it to act like diesel fuel, which has a much lower viscosity, it needed to be heated to more than $100°C$ to obtain that viscosity. The ship engine room is like a self-contained power plant and therefore the equipment to heat up the fuel could be installed easily, usually by steam coils in the fuel tanks. The heavy fuel also tended to have solid particulates and/or some water in it, and these needed to be removed as well, which was done by using centrifuges. However, heavy fuels are a complex mixture of hydrocarbons left after the more commercial parts have been recovered by distillation, so different sources of oil have different chemical make-ups. One consequence of this is that there can be some chemical reactions, particularly when one fuel has high asphaltenes and one has a paraffinic base, when a tar like solid is precipitated. This was found out the hard way on some cross-channel ferries, when a load of fuel was pumped aboard — or bunkered in marine parlance — in France instead of England. The centrifuges could not cope with the volume of solids being produced by the mixing of the fuel already in the tanks and the new French fuel. The lab was called in for rapid tests and established the high asphaltene content of the mixture. Thus, the report concluded: "Incompatible fuel bunkered." Or should it have been: "Incompatible fuel; bunkered"?

CHAPTER 3

The Wrong Sort of...

We explore how Scientific Services was involved when things went wrong because of snow, leaves and oil additives. Ian McEwen introduces the subject.

D URING A radio interview on 11th February 1991, Terry Worrall, the then Director of Operations at the British Railways Board, ascribed failures of certain types of electric trains to "problems with the type of snow, which is rare in the UK." The interviewer replied: "Oh, I see, it was the wrong kind of snow." Worrall replied: "No, it was a different type of snow." The Press had a field day and ever since has used the expression "the wrong sort of..." to describe weak or meaningless excuses for any sort of failure, from politics to space shuttles.

Scientific Services was used to dealing with the wrong sort of things, but usually the description was accurate — as it was for snow — and our role was not to excuse a failure but to remedy it, as the stories in this chapter show.

The Wrong Sort of Snow, by Dave Smith.

The phrase "wrong sort of snow" entered our psyche in 1991, when a senior railway manager tried to explain why snowfall had brought sections of the railways, mainly on electrified lines, to a virtual stand-still. The cold snap had been forecast and British Rail had claimed to be ready. Apparently, there are 35 types of snowflake, and the snow which fell at that time was not the norm for the UK. The snow was not deep enough (see photos 26 and 27) for snow ploughs or snow blowers and was unusually soft and powdery. It found its way into electrical systems, causing short circuits and traction motor damage in trains. For traction motors with integral cooling fans and air intakes pointing downwards, the problem was made worse as the air intakes sucked up the loose snow. Meanwhile, the snow also became packed into sliding door mechanisms and unheated railway points, causing them to fail. The disruption lasted for more than a week, causing delays of up to eight hours at a time and the necessity to replace several electric services with diesel traction.

Dr Peter Watson had recently been appointed as board member for engineering and was charged with sorting it out. Obviously, BR was incompetent, and no other railway would be so useless when faced with a bit of snow, so the government of the day decreed an answer! Peter was visiting the RTC in Derby a few weeks later and happened to spot me in a corridor as he toured the facility with the Director of Research. He had worked there more than 20 years before, and presumably remembered me from that time. He told the director there and then that I was the person he wanted to audit the railway operations for winter prepared-ness, and to prepare a report for the Minister of Transport on the cost of preventing future disasters, the Government having committed to invest in BR to avoid such disruptions in the future. This was a test for us in Scientific Services, as we relied on funding from all parts of BR — including track, signalling, train operators, design engineers and so on — and the prospect of assessing them and producing critical reports was indeed going to be, well, challenging.

Within a short time of commencing the task, I was at BR HQ with Peter, the then chairman of BR, Sir Bob Reid, and three other parties, where we discussed a plan of action. My prime task was to ensure the various businesses were supported in their winterisation response for the next winter, and to set up an audit programme for preparedness. There were other tasks that did not directly involve me. For example, Rod Smith, a learned professor from Sheffield University, was actioned to visit Japan to gain an understanding of how they prepared and dealt with snowfall.

Peter, meanwhile, was required to visit other European railway oper-ators to determine how they coped. He was taking a senior engineer from the Chief Mechanical and Electrical Engineers' Loco Design and Maintenance Department. The Government insisted that a civil servant should also be part of the task group. The chairman asked for the name of the assigned individual and when Peter disclosed it, Sir Bob said words to the effect of: "You don't want that prat with you! I will see what I can do." Duly, the trio departed to Denmark, France and Switzerland, and the discovery that Denmark suffered far greater problems than the UK was probably not news the Government wanted to hear. The civil servant did, however, feel able to make his point in Switzerland. He was aware, he said, that the Swiss rail operators had a fantastic chemical that when sprayed on snow could make it disappear almost immediately. There were some blank faces. He said he realised there were some health and safety hazards and that masks had to be provided to operators, but the benefits seemed to justify its use, according to a newspaper article in the Times. The Swiss representatives and UK officials made it clear they knew nothing about this process. The civil servant produced the relevant newspaper clipping which described the process apparently used in Switzerland to great effect... and the date of the publication was pointed out — 1st April! As you might imagine, the story flew around Europe and I believe was on the radio. Peter rang when he returned to tell me that the chairman was greatly amused! Unfortunately, the story never made the Press over here.

A key failure in times of heavy snow or other operational difficulties is the inability to adapt and communicate. Plans to ensure BR could operate skeleton services on key lines were formulated by all the businesses, and the matter of communication urgently needed to be addressed. This was very apparent on the journey home with Rod Smith, the Sheffield University professor. He was hoping to be home on time that night as there was an important Manchester United football match he wished to watch. The train we sat on at St. Pancras Station did not depart on time. After a five or ten-minute delay, an announcement was made to the effect of: "We are delaying our departure to take on the additional duties (stops) of the next train, which has been cancelled. I do not know the reasons as nobody has told me what is going on!" The journey was chaotic, and misinformation was given out sporadically without any real benefit to passengers. Rod was going to get home by midnight at best, never mind seeing the match. So that he was prepared, should he receive a phone call from Rod, the next morning I advised Peter Watson that the communication problem was still manifesting itself two months after the bad snow experience.

We greatly assisted Network South East, at their request, to improve their plans and they really got to grips with winter preparedness. However, in autumn the communication issues were still a significant problem. I was asked to attend an early morning meeting at Croydon; quite a task travelling from Derby with the "wrong sort of leaves" causing chaos at the time. When I tried to catch a Thameslink train at Kings Cross, there was utter chaos. The PA system sounded like it was ex-surplus supply from a submarine, with a guttural submariner making the announcements. Typical of the announcements being made was something like: "The next train on platform two is the delayed 7.02." As it was 8.30am, no one knew where it was going to. There were no illuminated signs at the time, so passengers rushed to the timetable to find out whether it stopped at their desired station. I was the first to arrive at Croydon by quite some time, as the local Network South East reps struggled to get to our meeting point. A report was submitted to the Secretary of State for Transport asking for about £280 million to prevent or minimise snow impacting on the system. Despite their previous commitment, the money was not forthcoming — a fact conveniently hidden by the furore at the proposed introduction and the subsequent withdrawal of the poll tax.

Scientists from all the labs — Crewe, Doncaster, Glasgow, London, Swindon and Derby — were involved in a massive exercise prior to the onset of winter in 1991. The aim was to establish if processes, resources and materials were in place should there be a repeat of the previous winter. On the civil engineering side, it was required to assess what they had in place for keeping the track open and the points free of obstruction by using heaters and special greases. On the operating side, an evaluation of stations and train operators was undertaken to determine if there was enough chemicals, grit or salt to keep stations operating and whether contingency plans were in place for implementing the emergency timetable. The operation of the trains necessitated that depots were equipped appropriately and gauzes available to fit over electrical gear. One of the successes in stopping snow entering the electric motors and other gear was to use women's tights. These proved to be a barrier to the very fine snow. Sir Bob mused whether he could use this piece of news to justify to the Minister that BR should open a Knickerbox at every station!

A report was provided to each business and the British Railways Board. Although many lessons had been learnt from the previous winter and from the preparedness audit, bad weather still seems to cause chaos and there is little long-term memory of what is needed.

The wrong sort of fuel, by Dave Smith.

There was no newspaper declaring "the wrong sort of fuel" but this story could have easily hit the headlines if the public had been aware of the facts. This story gives some insight into why we existed and why we were held in such high esteem by the railway industry. Our ability to take on major problems, make brave decisions and resolve issues quickly was our claim to fame. We were bold enough to challenge the oil companies about their technology and fuel quality. This is probably a story they would prefer to remain untold.

In the late 1980s there was complete chaos when about 50 Inter City trains came to rest through fuel starvation, caused by blocked fuel filters, on just one day — when the ambient temperature was just above freezing. The Mechanical and Electrical Engineering Department was the engineering heart of the BR train fleet and had a finger on the pulse as far as engineering matters were concerned. They asked for our assistance, as they considered it may be a fuel problem; BR uses gas oil in its diesel engines. Gas oil is, in simple terms, a slightly less well refined diesel than you find in the garage forecourt but is very similar in composition. At the time BR purchased about seven per cent of the total UK production for its diesel fleet, and Scientific Services tested the fuels as a matter of routine.

The pressure on us from the operating railways was enormous in resolving the problem quickly and permanently. Fortunately, we had an amazing arsenal of analytical equipment and talented scientists who were focused on problem solving and keeping the railways operating. Firstly, we needed to know what was causing the failure. Secondly, we needed to know the type of locomotives that failed and where these vehicles that had been refuelled. All this needed to be done within a few days if the system was not to come to a complete halt. No doubt the public blamed BR for the chaos but, as you will see, they were not at fault.

It turned out that most of the problems were associated with the high-speed trains which formed the backbone of the Inter City 125 services and some of the new Sprinter diesel multiple units providing local services. Locomotives had no morality at that time as they slept wherever they finished their travels, so tracking where they were refuelled was not as straightforward as one might think. The problem was made worse because fuel was bought from a number of refineries around the country, so there was not one unique supplier. Within hours, loads of smelly black bin liners arrived from all over the country with the failed fuel filters inside. We also received samples of fuel from the failed vehicles. These specimens came in vans and trains from all the depots servicing the failed diesel engines. Gas oil is very smelly, and it pervaded the air. There were oily bags everywhere and all we really needed were very small samples. The filters all had to be cut open to allow pieces to be removed. Then we needed to extract any debris from the filters to see if there was something unusual causing the blockage. Normally such failures are found to be some inorganic matter such as sand. This was quickly eliminated by our X-ray fluorescence instruments that showed the material was not inorganic. The preliminary work was carried out in our 'condom' laboratory. Condom seemed appropriate for a lab that tested oils and fuel — it being an abbreviation for **Cond**ition **O**il **M**onitoring (safe protection for your engine). Our analysis was able to determine the supplier of the fuels.

Our laboratory test on the blocked filters quickly established, via our analytical infrared specialist Ken Hart, that they were contaminated with an unusual polymeric material. What was a polymer doing in the oil? Was it a pollutant? Was it an additive? If it was an additive, what purpose did it serve? If it was an additive, it would be totally unexpected that it should block the filter. Remember all the fuels supplied were tested on a regular basis and conformed to the specified standard fuel clogging test (the Cold Filter Plugging Point test, or CFPP). The fuels were supposed to be operational down to -16°C. What was the source of the polymer? Although we were chemists, we were not generally privy to the wonderful alchemy that is used to formulate oils, paints and the like; such compositions are normally commercially confidential.

Most of the public are probably unaware that the diesel we buy at the pump in the winter is not of the same composition as the diesel we get in the summer. If it were the same, we would have major problems because summer fuel would form waxes and clog fuel filters, therefore starving the engine of fuel in the UK's cold winter months. The oil companies use additives to ensure oils do not wax in winter in normal conditions. For this reason, the composition of the fuel we use can vary from country to country depending on the climate and may change in those climates where the temperature drops below zero in winter. All the fuels tested by us met the specification and had met the required CFPP. So why did the diesels fail? If it was not resolved quickly, every diesel train on BR could fail and cause total chaos.

Together with the director of supply, we identified the source of the problem. The fuels supplied to the failed engines came from two oil companies and one specific refinery. The two companies, Total and Petrofina, were magnificent. They were most helpful and supportive and not at all defensive or obstructive. Without their cooperation we could have not resolved the problem so quickly. If only all organisations operated in such a positive fashion.

The other issue was the type of engines which had failed. They all had one common feature: they used fairly fine fuel filters which would be more susceptible to blockage from waxes. Although blockage by waxing is a known problem that can occur in severe weather conditions, the use of special additives in winter fuels helps to stop the heavy waxes coagulating and allow the engines to operate down to below -16°C. However, the temperature at the time of the failures was not even below freezing.

The first step then was to stop the affected fuels being used and replace them by supplies that did not appear to cause the problem. Although this was seen as a short-term solution, because failure to understand the problem may have led to a quick recurrence with other alternative suppliers, it was necessary. Had we not done so and made all companies aware of the problem, it is possible that the UK industry would have come to a halt. Total and Petrofina told us we had done a great service and we later became aware of the disaster we prevented: tankers would not be able to deliver fuel, trawlers at sea would be in great difficulty and eventually many trucks and buses would have randomly come to a halt.

John Sheldon, in charge of the oil laboratories at Derby, carried out special tests using filter paper with similar pore size to that used in the engine filters. We reasoned that this might give us some clue as to the nature and source of the problem. He found that fuels from the affected sources were very slow in passing through the filter at temperatures above 0°C. The deposits found on the laboratory filter was similar in composition to that found on the filters from the failed engines — a polymer. A laboratory test was subsequently devised by John that we called the cold flow test, which measured the time of flow of fuel at low temperatures. This was used for the evaluation of fuels and probably still is.

To save boring the reader, the convolutions of the investigation into why the fuel was slow in passing through the filter paper are not covered in detail, but the story is quite amazing, and the implications of our discoveries had significance way beyond the railway industry. In brief, the fuels are formulated by using particular cuts of the petroleum fractions that are produced at the refinery from crude oil, and then incorporating additives to produce a diesel fuel/gas oil that meets the requirements of the CFPP temperature criteria. This test was developed following extensive trials in Nordic countries many years ago. In effect, they discovered at what temperature vehicles could operate in practice and developed a laboratory test to reflect this condition. To prevent waxing in winter, additives were added to provide the winterisation properties they were seeking.

The fuel causing problems used a polymeric material to wrap around the waxes and stop them forming large conglomerates that could eventually lead to blocked filters. The resulting fuel passed the CFPP but no one, as far as we could determine, had tried the fuel in practice in actual cold weather conditions. The suspending agent used was the polymer we had found on the filters and was the cause of the blockage. To make matters worse, for this to work there was a need to add an additional wax. It may seem surprising, but this wax was needed to promote the process of suspending the wax by the polymer. This was not unique to Total and

47

Petrofina, and problems were identified to a lesser extent in most of the major suppliers. When the CFPP was developed the filters on engines were much coarser than those used in the modern high-performance engines of the day.

We worked extremely hard as a team, evaluating every facet of the fuel and additives used and the interaction with filters, pumps and so on before the fuel reached the combustion chamber. The problem was quickly resolved as a result, and the railways started functioning well again. Had we not resolved it, the issue would have become much greater, with the possibly detrimental additive packs becoming more common-place within fuel supplies.

Other organisations recognised that they may be facing problems. The Ministry of Defence and the blue light services particularly became concerned about the fuels they already had in stock and what might happen in an emergency. Such was the level of concern that we received several samples from them, but the level of security was so high that we were not made aware of the source of some of our samples. It was much later, when we were acting as technical advisor to the Leicester Health Authority, that the even wider implication of our work was realised. When they were asked about their winterisation provision, we were amazed to discover that their emergency standby generator fuel supplies were mainly in tanks stored above ground, and that they bought the fuels at any time of the year. Moreover, there was no programme for sequentially using the fuel oil, meaning some could stay in tanks for years. Storage over a long period was also a concern because the stability of the fuel was not perhaps as great as it should have been. The result of our tests showed that some of the tanks contained summer fuels and would not have necessarily worked in the generators during winter power cuts. You can imagine the implications for hospitals and what would have happened if we had a harsh winter. This resulted in all the emergencies supplies at hospitals being tested.

About this time, we had a visit from a Rail News reporter. We were looking to provide a story promoting our value within the industry, as well as outside. The newspaper had a limited circulation of about 150,000. The reporter had heard there was a potentially good story with respect to polychlorobiphenyls (PCBs) and the risk it posed to workers. However, we thought an article on this could cause unnecessary alarm. I, as Head of Analytical Services, suggested to the reporter that there was a better story that would give him a front-page headline and promote ourselves within the industry: the wrong type of fuel. It was obviously newsworthy, and many railwaymen would be interested to find out why everything had gone wrong very quickly over a few days, creating railway chaos. So, we told him about the fuel starvation but insisted that we vetted the article before publication. His report was a typical journalistic story told in a way as to attract the reader. Perhaps unusually, despite the language, we agreed for it to be published.

What was not good was for an edited version to appear in The Guardian and the Evening Standard, where the Head of Laboratory was attributed as stating that the diesel in the UK was a major problem. Perhaps unsurprisingly, he received phone calls the next day from two major oil companies making threats about court injunctions unless the articles were retracted. It was made clear that the stories came from an edited Rail News article, but we were happy for them to go ahead with the court action. They did not proceed. One of the major companies realised they had major problems and immediately started to respond. In fact, we advised our procurement department not to source other supplies of gas oil from one major source until we considered it met the new specification. The sad point was that as we were being privatised, committing the business to maintain the analytical facility to deal with such future emergencies was not there, should a similar problem occur.

The Wrong Sort of Leaves, by Dave Smith.

Since the phrase "wrong sort of snow" has been adopted and adapted by the Press and others to explain the demise of the 7.02 from Timbuktu or wherever, it was inevitable that the phrase would be applied to leaves on the line.

"I think I have spotted the wrong sort of leaf."

The wrong sort of leaves is by no means a new problem nor one restricted to Britain: they have been a recognised problem since the early days of railway operation. It was a problem experienced by Thomas the Tank Engine and his friends when he was created in the 1940s. A small team was set up in a Tribology Unit in 1967 to study wheel/rail adhesion and why trains sometimes lost traction. Tribology is the science and engineering of interacting surfaces in relative motion. It includes the study of friction, lubrication and wear; all of fundamental importance in railways. The word was coined by Peter Jost in 1966 in a government report which highlighted the high costs of friction and wear to UK industry. Thus, the Tribology Unit of BR Research was at the cutting edge of this new multidisciplinary science. The reasoning was sound — if the science and engineering were better understood, then solutions for improving traction might be identified.

The unit eventually brought together some eccentric characters who blended quite well to deal with the sticky, or perhaps not so sticky, subject of traction and wheel/rail adhesion. They were undoubtedly bright — many were young graduates — but very few could claim they did not manifest some sign of lunacy while working on this project. It required a mixture of skills to undertake research in a subject that interfaces with physics, chemistry, mathematics and engineering. The characters were quite unbelievable and kept most of those who worked there fully entertained by their idiosyncrasies. It was rather like being part of a Basil Fawlty sketch at times. In fact, the similarity was so great that Mick Broster, an amateur thespian, began writing a play for his drama group based on individuals in the office and the laboratory. Considering the success of the television series The Office, he should have persevered.

Most of this work to start with was laboratory-based (see photo 20), with the occasional foray to the railway line to take samples of the railhead contamination (see photo 21). The sampling element grew considerably with time. In 1969, it was brought to our attention that a train slid a considerable distance through a cutting at Sutton Bonington, Nottinghamshire, with its brakes applied. This was confirmed by mile-long marks on the track. The actual site was not very accessible, and the track was extremely dangerous from a working perspective. Trains travelled at 90mph and were upon you very quickly due to a series of bends in the area. The cutting meant there was little space to retreat to when a train bore down. The trains gave you less than ten seconds to move to a place of relative safety; a totally unacceptable situation in today's environment.

The cutting had many trees on both sides of the track, with steep walls for some considerable distance (see photos 18 and 19). On the Sutton Bonington side was a churchyard which became most eerie at night (yes, we had to work at night). The trees were heavily overgrown and hung over the track above the cutting wall for about 1km or more. In the autumn of 1969, leaf fall was particularly difficult. A gradual fall can probably be managed but a rapid fall, especially when accompanied by a number of other conditions, can quite remarkably affect adhesion.

Trains have steel wheels which run on a steel track. The contact between rail and track is important for several obvious reasons. The contact zone between a wheel and track is about the size of a one pence coin. As a train proceeds, it short-circuits the low voltage applied between the two running rails, setting both the preceding signal and the display on the signaller's screen to red, until the section is cleared. The presence of leaves means that, in the worst-case scenario, nobody knows where the train is because the leaves insulate the rail from the wheels. In these circumstances, trains can literally disappear from the signalman's screen. Because modern, lighter vehicles ride on the track much better than the vehicles of yesteryear, this is an even more significant problem.

Our interest was in the loss of traction and not the track circuit issue. Normal adhesion (coefficient of friction) levels of about 0.20 are sufficient for operating trains, even for those with many freight wagons attached. The presence of leafy deposits can lower this level to 0.01 or less. Trains cannot start, stop or even remain at rest on a gradient when the levels are as low as this. Such levels of adhesion cause absolute chaos and are quite frightening for the driver. He cannot go into a controlled skid or head for the field or the wall like a road vehicle; he just has to sit there and wait. The new vehicles now have ABS systems which certainly help but do not solve the low adhesion problems altogether.

The madness inherent among our ranks inevitably gave rise to unusual ideas. It was felt that we should consider all possible solutions to such problems, so we went to see Professor Wilkinson at the University of Nottingham's botany department, coincidentally based at Sutton Bonington. We asked about using Agent Orange, used with great effect to defoliate trees in Vietnam. This would have enabled us to get the leaves off in a controlled fashion. We reasoned that we could perhaps find something to bring them down on a Sunday and thus minimise the disruption. Apparently, the side effects of Agent Orange are grim (it kills people) so this idea was dropped. We did learn about aphids (greenfly to most of us) though. These are responsible for a leafy type problem in the spring when their sticky gum, sometimes called honeydew, falls from the trees in the cutting on to the rail. Should you wonder what that sticky stuff is all over your car when you've parked beneath a tree in spring, it may well be honeydew. If the wrong type of trees were about, the aphids had a field day, so honeydew was also a problem for wheel rail adhesion.

So — how do the leaves get on the railhead? At a meeting of high-powered senior railway officials, attended by members of the Tribology Unit, it became apparent that most had no understanding of the problems. Surely, they said, the leaves must be very deep on the track to cause such a problem. It must be very rare! Their ignorance and lack of awareness was incredible. How did these people run a railway?

We set about demonstrating what happened in practice. One of our most disastrous attempts was to film the movement of leaves. In those days, high-speed photography and video systems were not commonly available. We borrowed a high-speed camera from a company who were willing to demonstrate it to us. The first pictures showed Ian McEwen lying between the rails with a camera in hand, pretending to film as the train passed over him. In practice, as mad as we were, none of us would volunteer for such a task and we resorted to mounting the camera on a sleeper. It was a good decision because the film carousel was subject to impact when a train passed over it. The other problem to overcome was how to illuminate the underside of the train as it passed over the camera. We used a portable generator and some very high-powered lamps.

The representative from the camera company wanted to be present despite our warnings that this was not a good idea. The track had problems due to its foundations, giving rise to 'soft spots'. In such soft spots, the clay beneath the stabilising sand/stone layer is pumped to the surface by the movement of traffic over the sleepers, leading to track instability and poor vehicle ride. At the side of the track, the permanent way staff had made a trench which was filled with the slimy clay material removed from the pumped areas. We told the camera company rep to jump out of the way when a train appeared around the corner and so he did — right into the pit. He was in a right mess, or to be more precise, his suit was — but we still treated him to a nice lunch at the nearby Priest House Hotel. What they thought of our dishevelled, smelly guest with clay on his trousers up to knee height, one will never know.

The test itself was even more of a disaster. The camera worked, the leaves swirled as the train went by and leaves were deposited on the rails and trapped by the wheels. This we observed at a distance but failed to record on the film. As the train went over the illuminated areas, a passenger on the train used the toilet and the output hit the lamps, which exploded. We had a wonderful high-speed film of exploding light bulbs and effluvium!

We observed that fallen leaves on the ballast are sucked up and swirled in a vortex created by passing wheels. The air movement results in some of these leaves becoming trapped between the wheel and rail, and further traffic action produces a black deposit on the running surface. The imprint of the wheel is transferred again and again, and there is very little 'pure' leaf in the typical deposit on the railhead. The result is a tenacious shiny black film which can completely cover the running surface for more than a mile. So, all the falling leaves have the potential to get on the rails and signs of build-up are evident when only a few leaves have fallen. This situation occurs on dry days when leaves are particularly mobile. When it is wet, the deposit is not formed.

Our happy gang of tribologists spent many days visiting, measuring and sampling sites like the one at Sutton Bonington, involving lengthy trips throughout the country. However, we were asked to stop looking at leaves as the investigation into the low adhesion they caused was considered a seasonal issue and thus outside our real remit to understand the more general problem of low adhesion, which occurred from time to time throughout the year. But one year the leaf problem was so bad — especially on the Marylebone Aylesbury line — that we were encouraged to do more. I was required to travel on the multiple unit stock in the driver's cab on the line. We could do very little about the problem by riding up and down the line, but 'the powers that be' were insistent we attend as it showed commitment to the drivers; in other words, to demonstrate that BR senior management were taking the problem seriously. Sitting in the cab under such circumstances convinced me that this was indeed a major problem: at one station the train went backwards instead of forwards. As stated at the beginning of this book, the men and women of the railway made the system work often under very adverse conditions.

Why did the adhesion drop so much? With the very small contact zone between the steel wheel and the rail, trying to generate grip through leafy matter is obviously not ideal. However, the very low adhesion figures found were caused when the rails became wet after the dry leafy film deposit formed. Although the deposit did not form in the rain, when it became wet it lost all its structure and became a super lubricant.

The next question was how to stop it. The way the railways operated after the Beeching cuts and rationalisation led to many unanticipated problems by those seeking efficiency. This lack of knowledge of how things really function was a factor we recognised as the cause of issues and disasters over the years. Two things happened that impacted on this particular challenge. Firstly, the resource for maintaining embankments and the railways track was reduced, so no one was cutting the foliage that inevitably grows when it is not managed. Unsurprisingly, fast-growing broad leaf trees such as sycamore were recognised as a source of some of the heavy leaf fall. So, there is some truth in it being caused by "the wrong type of leaves". Our report and papers on the subject at the time (the early 1970s) were obviously not taken too seriously until more recent times. Maintaining the banks was a low priority and the message that this would lead to chaos in the future fell on deaf ears. It must, though, be acknowledged that many of the leaf-producing trees were outside of BR's control; often in domestic back gardens where the railway cuts a very narrow swathe through the surrounding area.

Secondly, as the name implies, a lengthman's job was to look after a length of track by regularly walking it to undertake minor repairs, report major problems and prevent the build-up of leaves or other debris. It was his domain and there were regular competitions to establish the 'best length'. The demise of the lengthman and the innovation of the mobile permanent way gang, introduced in the name of efficiency, did not give the same level of ownership and inevitably reduced the time for fringe tasks. So, the problem has got worse, and the newer lighter stock inevitably suffers more from poor adhesion and track circuit problems.

Research had been carried out many times to solve the problems of poor adhesion. Some of these remedies were well entrenched in the railway system. The most embedded one was the invention of Dr Andrews. He had a theory about 'secondary conditioning' (this is practically beyond explanation) and advocated the use of ethyl caprylate, a very smelly and sweet ester — and certainly a perfume you would not wish to wear. This was to be applied through a trackside device on those parts of the railway system experiencing frequent slipping problems. The device was not cheap and continually needed maintenance. It had a pump plunger which was activated every time a wheel passed over it, resulting in a spray of this extremely strongly smelling material. Having worked with it, the only comment you got when you travelled home was from some of your less cultured friends and family, who'd ask: "Have you been to a cheap brothel?" The smell clung to you and your clothing for a few days. As far as we could determine, it did nothing to the adhesion other than to reduce it slightly.

Some of our colleagues tried another approach, by applying sand in the form of toothpaste like extrusion onto the track. The material, called Slipmaster, was sand mixed with a polycell (wallpaper paste) type material and although there were slight benefits for the first wheels passing over the spot, the resulting mixture remaining on the track looked a bit like the dreaded leaf deposits. Moreover, it caused the leaf type problem again lowering adhesion downstream from the applicator. Meanwhile, we tried many approaches. We tested materials in the laboratory and the field. The amusing and amazing experiences while we did this are beyond belief–and none of the approaches worked very well or had a lasting effect.

We used high pressure water with additives. In the first instance we tried out some simple, fun tests in the laboratory. We used a motor to drive a standard 24-inch bike wheel in a specially built PVC cabinet. Inside the cabinet mounted to the wheel, we attached steel plates and bombarded them. The plates were treated to resemble the contaminated rail surfaces with radioactive tracer added. Then we sprayed them when they were rotating with various pressures of water, with and without additives. While these cleaned the surface to some extent and could raise adhesion, the problem was the practicality of application. To be effective the high-pressure water spray would only work when a unit was fitted to a train moving slowly. Trains moving at slow speed were also likely to bring the network to a grinding halt and were thus impractical. Some full-scale trials were carried out and the results showed the approach did not have a future.

Meanwhile, another part of BR Research constructed a mobile power station that generated enough power to operate a plasma torch. A train with the torches fitted travelled around the countryside causing chaos; it sometimes needed a fire engine to follow it. This was because, as an additional feature, it set fire to weeds on the track. It was frequently out on trial at night, so it did not interfere with normal traffic on busy main lines. This created another problem because the arc glowed very brightly and could be seen at a distance, apparently leading to phone calls to the emergency services and BR. Some callers thought aliens had landed!

We in the Tribology Unit were not directly involved in the development of the technology or the operation of this vehicle. We were unsure of the science: whether the energy dissipated was sufficient to remove the small amount of material at the wheel/rail interface (remembering that the rail is a huge heat sink). We gained some finance from a European fund for railway research to test out our theories. In the first place we contaminated pieces of rail with radioactively doped oils typical of what might be on the rail. Then we bombarded the small sections of rail with the plasma in the laboratory and counted the remaining concentration of Carbon 14. We concluded that the torch would have little effect on the surface of the rail and be unlikely in practice to have an effect operating at a few miles an hour. We never really considered what would happen to us from being exposed to a radioactive vapour! This outcome caused some grief as you might imagine. The plasma torch project was thought to have cost £500,000 in the early 1970s and represented a huge investment. The final blow came when we went to evaluate the train in a live situation.

The train was running through a leafy section of Bearsted Bank near Maidstone, Kent, early one November morning. It always seemed like we worked when it was cold and wet, and during very unsociable hours — and this was no different. The work necessitated an overnight stay in the Veglios Motel at Maidstone. The name I shall never forget. Having arrived on site we met the look-out appointed to keep us safe. A member of the local permanent way gang, he was of typical broad speaking Kentish stock. His first greeting was: "Where've you being staying, then?" When I mentioned the Veglios, he replied: "You mean the knocking shop; you'd be alright there then. I don't suppose you got much sleep with the all the comings and goings." His knowledge on this front was much better than his ability to keep us safe. He was supposed to tell us when a train was coming (well, I always thought that was what we had look-outs for) by blowing a horn and hopefully giving us time to get out of the way. Unfortunately, he was either blind or deaf or not very observant, because we had to pull him back from the track on several occasions. He was a liability, perhaps having spent too much time in the Veglios!

There was bad news: the adhesion levels on the track as measured by our portable plint tribometer (see photo 22), a friction measuring device, showed little improvement following the passage of the plasma torch train. I recall being hauled in by Dr Pankhurst, the Director of Chemical Research, when we produced a brief report of our findings. Things went very quiet, but this bit of work sealed the fate of the plasma torch train. This was most unfortunate as it had many knock-on effects for the Tribology Unit; it put an end to some of our trips out.

Four of us used to race around the countryside in a Morris Oxford Estate, jumping humpback bridges around Staffordshire and chasing up to north east Derbyshire attempting to measure the friction on normal (non-leaf contaminated) track before the plasma train came and then after it passed. Then we packed up and raced to the next site. The journey back was a little more circumspect. We would call at a farm at Pinxton and buy 36 dozen eggs, the occasional chicken and rabbit for our colleagues back at base. We did not make money on this — it was just a little service we performed. One of the criteria for being part of the team was to jump the humpback bridge at Whittington. You had to get the overloaded car to fly some distance on the other side of the bridge before you got your pilot's certificate. The car did not last too long. The scary part was driving down the M1 when we had a flat tyre. As we opened the tailgate to get the spare out, a 2.5-litre bottle of chloroform fell out and smashed. There was enough chloroform to put us all to sleep and it made changing the tyre an interesting experience.

The demise of the plasma torch train also ended other interesting sagas. They ran the train at night at a place called Bunny, very close to Gotham (we never did find Batman), in Nottinghamshire. Our party was in a Land Rover or some such vehicle while the train went backward and forward down the test track. We used a walkie-talkie system to communicate. Just before midnight we called the train and asked who wanted fish and chips from a nearby chippy because it was cold, and we were hungry. They gave us a lengthy order: pies, fish, chips, beans and more for the driver and crew, and the people on board who ran the vast amounts of equipment. There was quite a bit of ribald fun using the walkie-talkies as well: "This is Road Rover 2 over to Rail Rover 1." Not an unreasonable approach to life, I would hope you accept. But all our conversations were broadcast over the airways, and it apparently caused confusion over a wide area. A marshalling yard's operation became chaotic because of our interruptions to their system and we were duly reported. They were really using us to make the point that their systems were not good enough — but we still caught the blast.

This was not the first time we'd been in trouble through our night-time antics. We decided to test a theory that as the rail temperature lagged behind the ambient there would be the possibility of dew forming on the rail, causing low adhesion even on a dry day. This would explain the adhesion problems that seemed to occur when it was dry yet when drivers reported greasy rails. A gang of four or more of us sat out at night at a piece of track accessed from a small farm track near to Barton-un-der-Needwood, close to Burton-on-Trent. It was a remote spot with many other such tracks leading to unmanned crossings used by local farmers and railwaymen needing to access the track for maintenance. There were no immediate signs of civilisation and it was extremely dark. We took measurements roughly every half an hour or immediately after a train went by. This was one of the busiest lines in the country at that time, in the early 1970s, and may still be so. We needed a few of us: a look-out, someone to measure rail temperature, someone to measure the humidity, and someone to measure the adhesion and also to sample the track surface. We sometimes used a machine for sampling the rail surface contamination, which comprised of a bandage that moved between two rollers that wiped the surface. It was called the adhesion research sampling equipment (ARSE for short). Once we returned to the laboratory, we removed debris from the soiled bandage using a Heath-Robinson washing machine made of plastic bottles, elastic bands and a motor. Then we analysed the contents for oil and chemicals.

On this occasion we all worked very well, and the results proved our point. The work resulted in a published paper which was much referenced. Occasionally a train braked a little as it shot past but never thought much about it. However, we did notice a vehicle in the distance racing down a farm track parallel to ours. He was moving and then shooting back to the road again. There were many such tracks and it was evident he was getting nearer to us. This went on for a long time and soon enough, the vehicle disturbing our early-hours solitude arrived at our scene. It was a police car. The driver got out and stood with that confident look; you could almost see the sergeant stripes appearing on his arm. My supportive colleagues could see this was going to be difficult and amusing, and quite rightly decided that I should take the blame. They stood behind me sniggering.

The officer began questioning me: "What are you lot up to?" He had caught us red-handed, working legitimately. A train driver had reported us as "behaving suspiciously" and the stalwart officer had spent some considerable time trying to locate us felons! I explained that we were railway staff gainfully employed in our job. Although our dress would be more akin to felons out at night, I suppose our equipment and general demeanour did not quite fit in with the type classed as villains in his handbook. The officer was displeased to say the least. In fact, he was so angry that he removed his peaked cap and jumped on it. I had to keep a straight face and apologise for wasting his time, but I cannot say my rat-bag colleagues showed the same sympathy. They could hardly stand up! When the officer departed with crumpled muddy hat in hand, we all exploded into laughter. Thereafter, we always informed railway control of our activities.

The next such adventure was at Attenborough Station, near Nottingham. We went there shortly after the incident I've just recalled; in fact, I think we worked for about 36 hours or more without a break — totally unacceptable but loads of fun. This site had the advantage of having a railwayman's hut, and it was bliss. Half a dozen of us sat in this cramped little hut before a blazing fire. We brought bacon, sausages and the like, and we were having a good time playing cards. We had the odd drink too (non-alcoholic of course), and only occasionally was our enjoyment disturbed by a passing train. We knew when trains were supposed to pass and we tried to take measurements before and after the train, and at least every half an hour too, so we did work conscientiously, despite it disturbing some good card games.

What we had not spotted in our preliminary research was that one such passing train was the weed killer train. This swept the country spraying its cocktail of chemicals from tankers carried behind a locomotive. Of course, we were outside waiting for the train when this shot by. We were like drowned rats. "At least we can dry ourselves by the fire," said one of our number. Now this is when being a chemist is both useful and boring. I remembered that sodium chlorate was used for killing weeds and was a major component of the train's cocktail. This is a powerful oxidizing agent and so we realised that drying out by the fire might lead to us incinerating ourselves! We stayed outside, freezing.

It is also worthwhile mentioning 'The Misfits' as we called them and our exploits at Barton-under-Needwood. We would go to the permanent way gang's hut at what was the old Walton Station area fairly regularly, building up a picture of adhesion, applying our mixtures and trying to establish what benefits there were, if any. The hut frequently contained the members of the gang that our lot christened 'The Misfits'. We were fairly young and not so worldly in some of our experiences as a team. None of us had quite ever seen such a bunch. Some were friendly but others looked like they were on probation — and probably were. They always seemed to be there, no matter what time it was. Tea was always brewing, and the drawers and cupboards full of porno magazines to pass the time away. We got braver as time went by and asked: "What's up" and other such local vernacular. They said they did not work when it rained and there was bad visibility. On the days when we arrived and it was dry, the response was: "It looks like rain — we will see what it does." One of the general track gang rules was: "In rain and falling snow, into the hut we must go!"

They gave good advice, though. "You see the armband you have on for being a look-out?" they said. "Well should anyone have an accident, remember to look for their left arm and put the badge on it. You can't bring them back, but it stops lots of enquiries for the survivors when there is an accident." If a look-out is killed, who is there to blame? Whether there is much truth in such stories is not known, but we did notice that look-outs were often an older and not-so-stable men. There was some sense in this, but they were usually pretty bad as look-outs because it was important to have good eyesight and hearing. They were tested upon joining BR and retested as they got older, but it would seem they somehow passed — but could hear nor see trains.

There were more advantages at this site. It was local and the track was straight, so we could see the trains coming. It gave us plenty of time to move equipment out of the way — and ourselves. It also had the benefit of not wasting too much time travelling should the weather change. We would work when it was dry and then apply water or chemicals to assess the effect of our treatment. If there was doubt about the weather, we would ring Watnall Weather Station in Nottingham, a few miles from Derby, and ask for their forecast. One day the forecaster said it was going to be dry. We asked: "Well, when will it be dry?" The response was: "All day." We replied: "It is raining here in Derby." "Is it?" the forecaster asked. Our faith in weathermen was never quite the same again.

The visits to Walton were fun. At lunchtime we went to a local pub which was frequented by the gentry and not the likes of us in jeans and donkey jackets. The landlord always made us feel unwelcome, so we persisted in calling. He had good food and some very nice relish — we always emptied the jar of mango chutney. Otherwise we went to Lichfield where there was a super Chinese restaurant with a very favourably priced three-course lunch. We would then go to a pancake café afterwards to see who — after a big breakfast and the three courses — could down most.

Some days the team went off early and had breakfast on the way to the selected site. The lads liked the most unlikely places. Admittedly the expenses were not great but the places we visited would beggar belief to many observers. A transport café located strategically near Burnaston Airport (now Toyota), was the ultimate greasy spoon. Our server suited the place. His apron was the same week in and week out, and the long stalactites on the ketchup bottles said it all about the place. It would have been a suitable training ground for any environmental health officer! The food was not bad though. Mick Broster, by far the smallest of the team, marched in one day and surveyed the menu on the blackboard. "I will have that please," he said. Our man behind the counter, not known for his fine use of the English language, replied: "What do yer want?" Mick said: "All of it. I am hungry today." The owner of this most salubrious eatery had obviously not received such a request before. He was probably also surprised by our Egyptian colleague who ordered a bacon cob. We told Fuad, a devout Muslim: "But bacon is pork." His response: "Bacon is not like real pork and it is all right for me to eat."

Fuad was most interesting in many aspects of his life. One the managerial chores was stopping him chasing anything in a skirt while he was at work. His exploits with women were a nightmare. He was trying to get funds for staying at university and this required servicing a Jewish lady at the time when Arabs and Jews were not getting on too well. Upon mentioning this to see if it was an issue, it was really rather like his attitude to bacon: "If it tastes nice..." He also fell foul of the police, who stopped him and found he was not in possession of a tax disc, MOT or whatever. A few weeks later he was stopped again. He told them: "You cannot prosecute me for this as you did me a few weeks ago."

Of course, we occasionally worked!

Ian McEwen, also part of the team, here adds to Dave's memories of this time. We have heard about about leaves on the line and their influence in reducing adhesion to very low levels, particularly when damp, he writes. However, the profound effect really had to be seen to be believed, and the work at Bearsted Bank demonstrated it frighteningly. The site was well-known for low adhesion incidents in the leaf fall season. I was a member of the track team on the ground, with others on a test train hauled by an electric loco. At the end of our work the locomotive was to stop at our location so we could all climb aboard with our equipment. We saw the advancing train slow down to crawling speed but quickly realised it was not going to stop. As it came close, we could see that all the wheels were locked, and the entire train slid silently past us on the downward incline. The driver had no control whatsoever. Because of the quietness of the electric locomotive, it sounded to us like skaters on ice gliding past: a sort of shhhhh noise. There was a continuous black film on the rail at the time. When damp, leaf films can produce the lowest levels of adhesion measured and yet there are still those who regard leaves on the line as a joke or an excuse for bad driving. That afternoon taught us that we really had to find a remedy for this situation.

In the early days of track testing with the prototype Advanced Passenger Train (APT), Slipmaster was applied to the rails in braking tests. The purpose was to ensure a high enough adhesion level between wheel and rail to allow the train to stop safely from high speeds. In fact, it was found that the material could, in certain circumstances, have the exact opposite effect and produce slippery rails, so the Tribology Unit was asked to investigate. As previously explained, Slipmaster was a gel with the consistency of toothpaste which contained sand for adhesion improvement. The gel was achieved using two organic compounds akin to wallpaper paste and was intended to hold the sand to the rail it was applied to. Tests with the tribometer train demonstrated that, as wheels passed over the treated rails, the sand was crushed until the properties of the gel base became more influential and reduced the adhesion. After the passage of only 12 axles or so, the measured adhesion started to fall. This was confirmed in small-scale Amsler roller rig tests in the laboratory, and the search began for an inorganic gelling agent which might not reduce adhesion.

A material called Laponite, a man-made clay material in fine powder form, was featured on the BBC programme Tomorrow's World and appeared to hold promise. Laboratory tests began to reformulate a sand gel with good adhesion properties. In order to disperse the powder in water, a high shear rate was required during mixing and our initial formulation tests were carried out in a domestic Kenwood Chef liquidiser. Using an 1.8 per cent dispersion of Laponite in water gave stable clear gels which were capable of supporting 30 per cent sand and were easily pumped. We called this product Sandite (sand and Laponite), a name still in use today. Most importantly, laboratory tests showed Sandite displayed no adverse effects on adhesion and potentially improved it, even when the sand was crushed in the wheel/rail contact. This was then confirmed in a series of field trials using the tribometer train.

It was obvious that Sandite offered a remedy for situations of low adhesion on the rails, and development work proceeded with this in mind (see photos 23, 24, 25 and 28). It was thought that Sandite could have use at sites known to suffer from seasonal leaf fall where adhesion could be exceedingly low, affecting both traction and braking of rolling stock. For site testing we moved up from the kitchen liquidiser to powerful paint stirrers capable of dispersing the powder for the gel in large drums, and eventually to a cement mixer to make larger batches.

In 1975 and 1976 we tested Sandite as an adhesion improver on the South Devon Banks where the tree-lined inclines had caused problems for many years. Not only did the low adhesion produced by leaf films on the rail affect train running, but the rail burns produced when wheels slipped or slid were costly in terms of repair or replacement. In October 1975 a team of three chemists put the cement mixer, a 60-litre storage tank and other equipment in the guard's van of an HST leaving Derby, knowing that the unit would terminate at Plymouth and proceed to the Laira depot for servicing. We off-loaded our equipment and returned to the depot the following day to begin mixing Sandite batches. We planned to apply the Sandite from a Class 25 loco, using a peristaltic pump to supply nozzles held just above each rail. The tribometer train measured adhesion levels before and after application of Sandite, and at times to fit in with timetabled traffic movements to test the lasting effect, if any, of our treatment.

On day one we were given a corner of the depot with power and water supplies and, in our white laboratory coats, caused quite a stir with the maintenance staff. People stopped by to ask what we were doing, and we gave the spiel many times over that morning about mixing this wonder gel to treat the rail. We had just one problem. We set up the cement mixer, added the water and then the calculated amount of powder and...

nothing happened. No gel. News got around that the boffins were having issues and several men came over to look at the water sloshing about in the mixer. We left it to mix for a while, but still no gelling. We stopped the mixer for a while — nothing. We added extra gelling agent — still nothing.

We did the most logical thing and stopped for a cup of tea. We had stayed the previous night at a B&B and at breakfast, someone had commented on how good the tea was. As we puzzled things out, our most junior team member said: "Well, the water might make good tea but it's no good for making Sandite." And the penny dropped! A couple of years before I'd had an interview for a job with Acheson, a company that made colloidal graphite. They said one reason they chose Plymouth for their manufacturing base was because the water was so soft. I then remembered that gel formation could be aided by the presence of electrolytes in the water, so I dispatched our junior member to the shop for a bag of table salt. We had a bit of an audience at this stage, and a fair bit of sniggering. However, we turned on the mixer, added 10g of salt and bingo — within a minute we had a thick gel suitable for adding the sand. A little bit of chemistry and sprinkle of table salt had saved face, and our tests went ahead successfully.

Sandite is still one of the most effective remedies for low adhesion and has been produced commercially in large volumes for many years, with formulation changes along the way to improve storage stability, anti-freeze performance and so on. A product containing stainless steel particles is used to improve track circuitry. It is now used in Ireland, Holland and Belgium, as well as the US. There is a great deal of satisfaction in solving a problem with a very simple but elegant solution, and salt was not the only grocery item that the inventive tribology team used to further the progress of science on the railway.

In the early days of measuring wheel/rail adhesion, the tribometer train was developed to measure the 'true' adhesion between a full-scale wheel and the rail on any stretch of track. In essence a two-axled goods van was adapted and instrumented as such that each axle could be braked independently and the measured forces at the point of slip could be computed to give a coefficient of adhesion and the brake force immediately 'dumped' (meaning dropped to zero). The train could be run at most line speeds and alternating axles braked to produce an adhesion profile along the track. We also had a portable tribometer that a team of two to three people (plus a look-out!) could use to make measurements at any reasonably accessible track location. The planning needed to allow the movement of the tribometer train, and to gain access to running track was difficult and had to be done well ahead. The portable tribometer could be used at short notice and did not require a possession to be used. A series of tests were arranged to compare the results of the full-scale and small-scale tribometers so correction factors, if needed, could be applied to all future measurements. A set of test routes were chosen so measurements made on live track with the tribometer train could be followed up as quickly as possible with measurements from a team moving around by road to use the small-scale equipment.

We needed to get as close as possible to making the two sets of measurements at the same position along the rails, and we needed a method of pinpointing where the measurements started and ended. This was before the days of accurate GPS equipment and although we might start testing at, say, a defined milepost, we thought it preferable to mark the locations when the tribometer vehicle started to gather results. We considered a paint spray gun that would squirt into the cess at the side of the track, but at speed this would spray only a thin film of paint, probably over several metres of track. Paint balling guns were unheard of at the time.

I have no idea who thought of it, but our chosen solution was very simple, cheap and required no equipment. We bought from a supermarket a pallet of 1lb bags of flour (I can't remember if it was plain or self-raising). As the tribometer vehicle ramped up the brake force, we dropped a bag of flour from the window of the adjacent coach on to the cess to mark the spot. The paper bag burst on impact with the ballast, producing a white marker which could be seen from quite a distance by the track team. They would then make portable tribometer measurements in that same position. Even if the track team took some time to get to site, following the train by road, the marker was visible for several hours. Over the next few days, wind, passing traffic or rain dispersed the flour, leaving no significant environmental impact. And in most cases, the track team picked up the torn paper bag and removed it from site. A cheap and simple solution, if not exactly elegant.

We have seen how leaf films on the railhead can have a profound adverse effect on wheel/rail adhesion. But leaves are not the only contaminant found to affect adhesion and in nearly every case, the result is to lower adhesion. We investigated a location where reports from drivers were of greasy rails. When we arrived, it was obvious that the adjacent door manufacturer premises had a problem with their extraction equipment. Sawdust was being blown on the wind and on wet days it was sticking to the top of the rail. The subsequent passage of train wheels was producing a film very much akin to leaf films.

This type of contamination from industrial properties adjacent to the track is not uncommon. In Scandinavia there are reports of pine needles getting on to the rail and also of tree sap being transferred by aphids. In Japan at a certain time of year, migration of a sort of worm results in them getting caught under train wheels, and similarly with locust swarms in Africa. All these phenomena effect adhesion adversely.

Sometimes, however, the train drivers themselves come up with imaginative reasons for their problems with greasy rails. We investigated a site adjacent to RAF Cosford, where the end of the runway was a very short distance from the railway track, the two being at 90 degrees to each other. The story gaining management attention from the drivers was that planes coming in to land low over the railway line were dumping fuel to make an easier landing, and that aviation fuel on the rails was lowering adhesion. We spoke to an RAF station officer who dismissed this as nonsense, but we visited the site to find out why the adhesion was low. There was no evidence of aviation fuel contamination and a chat with some of the allotment owners on the opposite side of the track established that they were happy eating their vegetables and had no complaints about RAF flights. We did, however, ascertain that a short distance away in each direction were trees which could undoubtedly cause problems in the autumn due to leaf carry-over.

Ian Cotter, then at the Swindon Area Laboratory, recounts an accident due to leaves on the line which illustrates how everything should be considered during the introduction of new vehicles and technology to the railway — and shows how easy it is to slip up. The Swindon lab covered the Western Region, offering all sorts of scientific services to the railway. The lubrication and wear and tribology teams, through Ian McEwen, undertook widespread trials of wheel/rail adhesion. They developed methods of checking for materials that might be undesirably lubricating the rail surface, causing wheels to slip both during acceleration and, more importantly, braking. The general method of braking on BR was mechanically pressing a cast iron block on to the wheel rim, a method which started with wooden brake blocks in Stephenson's locomotives such as the Rocket in the mid-1800s. In the 1990s new trains were being introduced with an innovative system called air-operated disc brakes. On the Western Region, disc brakes were fitted to the lightweight Thames Turbo DMUS, with newfangled electronic wheel slide protection which automatically took control from the driver, and suspension that made the bogies ride steadier on the rails. The old trains had leaf spring bogies, cast iron brake blocks and no electronics at all. These new features caused an accident at Slough on 2nd November 1994, which Ian Cotter describes for us here.

On the fateful day it was dark and raining as the 19:41 Paddington to Slough three-car Thames Turbo train 2F76 left the station. The rain turned to drizzle during the journey and at Langley, the station before Slough, there were five passengers and the driver on board. There is a downward gradient towards Slough, and when the driver applied the brake in the normal fashion it failed to have much effect. The train was sliding with the automatic wheel slide protection system operating, and the driver had no control over this. Even though it was possible that, with his experience, he would have been able to reduce the speed of the train, having no phone or radio meant he had no communication with anyone outside the unit. Thus, the signalmen were oblivious to his predicament. A full emergency brake application had only a small effect, reducing the train speed from 56mph to about 36mph, as the wheel slide protection still operated to constantly apply the brake and then release it as the wheels slipped. At Slough, the train was signalled into the bay platform — a dead end with a buffer stop — and the inevitable happened.

The sliding train passed over the crossover to the bay platform at more than twice the permitted speed of 15mph. The driver, anticipating the inevitable collision, lay on the floor of the passenger compartment. The train hit the buffer stop at about 36mph and ploughed through it, over the platform and into the booking office. The driver and a member of the station staff were sent to hospital, thankfully with only minor injuries, and no passengers were hurt.

The catastrophic loss of adhesion led to the head of the Swindon lab to check the rails for contaminants, arriving on site at 1am on 3rd November. The initial inspection revealed what looked like debris from leaves on the line, with a small amount of water on both rails over the whole of the length of the slide. Her Majesty's Railway Inspector convened a meeting in the early hours and was informed of the initial inspection. Extensive laboratory analysis was undertaken to confirm leaf debris and where it was situated. Leaves were found all the way from Langley station to the bay platform at Slough. The area that the leaves had dropped on was before Langley, some two miles from Slough. The debris was carried all that way by the new train wheels because they had better suspension, allowing the wheels to travel more steadily on the rail. Older trains 'hunted' from side to side, thus scrubbing the rails, and there were no brake blocks bearing on the wheels to scrub them clean. The drizzle had had just the right amount of water on the rails to activate the debris into a thin paste that was an excellent lubricant. No one had done an effective evaluation of the new trains with respect to the adhesion issue.

CHAPTER 4

Helping The Police With Their Inquiries

Here, with the help of Vince Morris, we discover the involvement of Scientific Services in crime detection and prevention, from the Isle of Wight to Silicon Valley.

THE OFT-HEARD statement that the British Transport Police (BTP) are "just railway police, not real policemen" has led to the downfall of many criminals, since the BTP has the rights and responsibilities of other police forces. Indeed, they — or rather their predecessors employed by the individual railway companies — were pioneers in many fields, including the use of dogs and the employment of women as fully sworn constables. They were also early users of forensic science, much of which, even in the 19th century, was undertaken on their behalf by the railway chemist. Today the BTP is recognised as an extremely efficient force in the forefront of anti-terrorist activities as well as having special expertise in the management of major incidents and crowd control.

In the period covered by this book, forensic science was maturing into the activity we know today (but still very different from its TV representation). To put such advances into perspective, consider the following. The personal computer was introduced in the 1980s, mobile phones (the size and weight of a brick) were introduced in 1983, DNA was first used as a forensic tool in 1986 and although BR first used CCTV in 1965, it was not installed as a permanent fixture on the railway until 1975, when it was introduced as an aid to drivers on London Transport (it is said that the first cameras were regularly stolen!). Credit cards were widely introduced in the UK by Barclays Bank in 1966, joined by the Access card in 1972. Instrumental chemistry was in its infancy but greatly boosted on BR to the envy of many commercial laboratories when, in 1965, the Central Analytical Unit opened in Derby, and shiny new instruments replaced a lot of the wet (or 'test tube') chemistry. Even in the early 1970s we were using electro-mechanical calculators which, in one area laboratory at least, were chained to the desk in the calculation room; not against theft but to ensure they were always in their designated place.

In the 1960s, most of the forensic work covered by the BR chemist was outlined by Eric Henley, then Deputy Director of the Scientific Services, as guest speaker at a sergeants' course held at the BTP training school in October 1967. Topics covered included document examination, oil on navigable waters, dirt, dust and debris, and fire and tool marks. By the 1970s, Scientific Services staff, mainly from the London laboratory but also Glasgow and Derby, were regularly lecturing at the training school, and cadets were seconded, at the age of 18, to Scientific Services for a four-week period. This was typically made up of three weeks at the London laboratory where hands-on work was encouraged, and a week at Derby where the wider aspects of BR Research were demonstrated (see photo 31).

So, what sort of cases were being handled? The following contributions may give some idea of the jobs we were involved in, bearing in mind such was our arrogance in our own ability that we were loath to turn anything away. We were accepted in court as experts: the expertise of a forensic witness is, in theory, decided by the judge, and they decide whether the evidence given should be admissible in their court (see photo 29). They can use the well-worn definition of an expert, that an 'ex' is a has-been and a 'spurt' is a drip under pressure, or a more suitable one like 'an expert is someone who, through training, qualification or experience can tell the court something which the court would not, of itself, know'. We never knew which one an individual judge used in our cases. In the early days we received no formal training in giving evidence, but assumed that one should look as smart as possible. I fell foul of this on one occasion when I turned up at the lab dressed to visit an arson scene in a T-shirt, jeans and a cord jacket (all the rage) — and then got an urgent phone call saying I was required at court in Knightsbridge where a case was not going as expected and was being delayed until my arrival. I arrived in what I considered to be very inappropriate dress for an 'expert' but not a word was said and my evidence was accepted (see photo 30). The only other convention I did not follow was that I had a beard. A solicitor told me people with beards were less likely to be believed than the clean shaven (I still wonder if that was true). I had a beard because my wife told me no policeman would listen to me as I had a baby face. I have been hirsute ever since, but now it is just laziness. Here, Dave Smith tells us about forensic work that could be fitted into the laboratory routine.

A routine job?

The most routine job, writes Dave, was probably the examination of oil in navigable waters, comparing samples extracted from polluted water courses and harbours with samples taken from the suspect vessel's bilges (at some risk if the captain was not too co-operative). Since a major part of the every-day role of the laboratory was to use instrumental techniques to determine the composition of engine oils, this was just one more analysis. It rarely resulted in a court appearance since the vessel had sailed away before the results were obtained!

Another regular area for forensic analysis was metal thefts. BR was always chasing those who stole railway assets; such thefts had a major adverse effect on the operational railway. Axle boxes on rolling stock contained high-value brass bearings. It was fairly common practice for some of the less benevolent members of society to jack up wagons and remove the bearings, and then replace them with the wooden chocks used for rail fastenings. The vehicles would move off and at some point in the journey, the bearings overheated and fires broke out. No doubt the delays and derailments that occurred were blamed on the incompetence of BR rather than the crooks. There was a factor in our favour. Anyone handling such bearings inevitably got their clothing contaminated with the lubricating oil. This material was unusual in that it contained ten per cent rapeseed oil and our analytical techniques allowed us to determine a component called erucic acid. Many a rogue was successfully prosecuted on this evidence (see photos 32 to 36).

Another lucrative trade was to steal the overhead telephone cables to signal boxes and stations that ran alongside the track. This was high-value copper wire. In a case from North Derbyshire a farmer reported unusual activity alongside his land adjoining the railway line. Two men were acting suspiciously: they looked like railway men but were working far too hard and too fast! No regular police officers were available, so the dog handler was dispatched from Buxton. He arrived and found two men sunbathing on the back of a trailer. Crime solving was possibly not one of his strong points, as he asked the men whether they had seen anything suspicious. They told him no, of course. It was only as he walked away and tripped over rolls of stolen cable that it occurred to him that these two might actually be the villains — as subsequently proved to be the case when the metals were analysed by Heinz Bauer, a very thoughtful and enthusiastic scientist who loved playing detective.

But there were other, usually one-off jobs. Here Vince Morris lists a few in which he was involved, mostly from his days at the London laboratory.

How safe is a power station?

The power station at Norwich was near the station, adjacent to the current Crown Point depot, and had rail links for the delivery of coal. When it was being demolished, the contract included the removal, for scrap, of the internal railway system. On-site demolition staff took the concept of removing the railway a bit too literally and did not seem minded to stop at the end of the Central Electricity Generating Board (CEGB) track — they were suspected of lifting BR track too. But was this a local enterprise by one demolition gang, or a company policy? We were asked whether it was possible to determine if any rails in a particular load leaving the site by lorry were of BR or CEGB origin. It was doubtful whether this could be done but there may possibly have been a metallurgical difference in the rails if they were sourced at different times from different suppliers. It was worth a visit to take small swarf (particulates generated from drilling) samples for chemical analysis by drilling into the rails on the back of the lorries as they left the site.

It was, as suspected, ultimately fruitless. But the main reason this particular job was memorable was that the officer in the case suggested that while the chemist sampled the rails, he himself would stand well back from the partially demolished building. After all, he said, demolition was a dangerous process. Only after the samples had been obtained did he further explain that it was well known at Norwich for debris to inexplicably fall from great heights onto people sniffing around the area, especially if they were accompanied by uniformed police officers.

All at sea

Contractors were employed to repair the road surface at the Isle of Wight car ferry landing terminal at Fishbourne. Since the roll-on roll-off ferries lower their bow ramps heavily onto the terminal decking, the specification called for a strong bitumen road surface on the concrete base. After a very short time the surface began to fail, and it was suspected that the material used was not to specification. The contractor denied this, so I was despatched to the Isle of Wight to take a sample. Once back in the laboratory, dissolving the bitumen to determine the stone content in the surfacing was a straightforward job and it revealed, as expected, that the material was not as specified. The job itself was quite simple but the instructions I had to follow to get to Fishbourne were more complex! I had to get a train from Waterloo to Portsmouth Harbour Station (which connects with the passenger-only ferry to Ryde) then "leave the station on the seaward side (there is no public exit — just do it) and shout for the Sealink tender which will be in the area, and ask them to take you across to the car ferry terminal on the other side of the harbour". If memory serves me right (it probably doesn't), the tender was called Sally, so I was standing on the harbour calling out "come here Sally", and remarkably it did — but it did not stop. I had to jump on as it went past the steps, state my destination and jump off as it passed the steps at the other side. The crew did not ask who I was or question why I needed to go to the car ferry terminal. For the local staff I am sure it was an everyday occurrence; to me it was another experience in the life of a jobbing forensic scientist.

Blacking out Suffolk

Firearms did not fall within our expertise, but the examination of a crossbow was something I conceitedly considered within our abilities. It could, I reasoned, always be sent to the aerodynamics laboratory in Derby if it became a bit difficult; bullet trains in Japan, crossbows in Suffolk — not much difference. Trains on the East Suffolk branch were arriving in Ipswich with severely dented side panels. Observations by BTP established that a local man was firing a crossbow in the wood beside the track, but when detained he stated that he was hunting rabbits and would never fire his bolts at a train. I was asked to examine the crossbow, safely stored under lock and key at Ipswich Police Office. It was home-made, using a wooden chair arm as the stock, and he was, indeed, firing bolts: six-inch ones with nuts still attached! Could such bolts cause the damage on the trains? There was a good mechanical fit but the distance the bolt would travel could not be determined, nor what damage it would cause to a steel panel. I mumbled something about taking the bow to Derby, but the obvious thing was to go to Suffolk Constabulary and ask a firearms officer to test-fire the crossbow at their shooting range. This they readily agreed to; they did not get a chance to fire crossbows very frequently.

The BTP officer in the case and I went into the basement of Suffolk Constabulary HQ where their shooting range was housed, and handed the home-made crossbow to the firearms officer. The range was set up with targets lit with spotlights and protected by heavy metal shields to prevent stray bullets hitting them en route to the target. The officer took the bow, placed the bolt on the stock, pulled back the wire of the firing mechanism, took aim, fired… and blacked out the entire range as the bolt followed an unexpected trajectory. It swooped towards the floor, then rose as if to return towards the firing position, but instead went behind the metal shield and destroyed at least some of the spotlights, blowing the fuses. It was difficult to gauge who was the more embarrassed; the firearms officer for the wayward shot or the BTP officer for introducing such a weapon onto the firing range. It never was determined whether a bolt from the crossbow could have caused the damage to the trains but we had little doubt that the owner had no idea where his bolts would go, and no more incidents were reported after his bow was seized.

Have yer gorra a light, boy?

Sometimes work involved helping to establish what a defendant should be charged with. Another job in Suffolk involved an attempt to set fire to a train. These were the days when oil lanterns were used to protect roadworks overnight. Like the signal lamps used on BR, these lanterns used a form of kerosene and a lighted wick and, being unattended, they were a target for the local youth to play with. One such group thought it a good idea to take a lantern and pour the fuel onto the roof of a train as it passed under a road bridge, and then rush to the other side of the bridge to drop burning matches. When apprehended, they said they assumed it was petrol and that a spectacular blaze would result. Kerosene is far less easy to ignite, requiring a wick, so their efforts were doomed. To confirm that it would not be possible to get a good blaze going, an attempt to replicate the process was made in the laboratory, or rather, outside it. The matches went out during their descent, so the burning properties of the fuel were irrelevant. Should the charge be criminal damage, attempted arson or even endangering the lives of passengers? Luckily this could be left to the legal department to decide, but they were supplied with the information that the youths' intention would have never been satisfied.

No accounting for Cheddar

Some crimes are simple to understand but for a successful prosecution, all angles must be covered, and this involves the use of experts. It was in the early 1970s that a meeting was held at Victoria Police Office with a detective constable, a BR accountant and myself, accompanied by a new recruit to the London laboratory, Arne Bale, who I was mentoring. The reason for the meeting was to discuss 'the great Cheddar cheese fraud'. In those days, Travellers Fare — the catchy title for BR's catering department — made its own sandwiches for sale on trains and in buffets. Like most catering organisations, demand fluctuated daily, so the amount of cheese required could not be easily predicted. To overcome this, the wholesaler employed drivers who visited the various food outlets and supply cheese on demand, making out an order in a book signed by both driver and caterer. Typically, the driver was paid in cash and the amount he paid in on return to the depot matched the amount of cheese he sold. Most eateries are small family concerns and the same person would order, accept, sign and pay for the cheese. However, BR was never that simple. The sandwich maker was not in the loop: his boss ordered the cheese but did not handle cash. In fact, no one handled cash. The order had to go to the accounts department, which knew nothing about cheese but would respond to an invoice. This process took several days so it was difficult to tie up the actual cheese, probably already consumed by passengers, with an invoice arriving in an office in Croydon.

"I smell a rat!"

All these complications made the paperwork very different from the van man's typical customers. Indeed, BR produced its own order paperwork which the driver had to use. This involved several carbon copies of the original order made out by the catering manager. From memory there were seven, including the top sheet. One stayed with the cheese, one was filed by the local catering department, another went to Travellers Fare HQ and one was sent to the BR accounts department. The van driver kept one, another was kept by his depot and one was sent to their own accounts department to allow an invoice for BR to be raised. If the copies did not correspond to the actual order, payment would not reflect the amount of cheese actually sold. So if the order kept locally indicated, correctly, that the driver had delivered 20lb of Cheddar at a cost of, say, £20 but the invoice sent to BR said the delivery was for 40lb, the cheese wholesaler would send an invoice for £40 and the total amount of cheese apparently sold by the van man would be reflected in how much money was raised from the total deliveries.

But the van man would have 20lb of cheese, to be paid for by BR, which he could sell as a local enterprise to the next customer, destroying all the associated paperwork so his bosses would be no wiser. The fraud would only work if there was collusion between the delivery man and the person on the ground at Travellers Fare making out the order and receiving the cheese. In whatever way the financial split was made, both would finish quite a bit richer, and if the relevant Travellers Fare man was not around on a particular day, the delivery could be made legitimately with no need to compromise the scheme.

To make sure that the fraud worked, it was necessary to ensure that paperwork was suitably doctored to allow the appropriate copy to go to the appropriate departments. This involved inserting cardboard above sheets where you did not want a copy to be made, then going back and making incorrect copies to show the wrong weight of cheese and payment due. This was where I came in. Could I examine the paperwork, all seven copies now being united by the BTP, and determine whether cardboard had been inserted, and show that not all the copies were identical? Although the DC could easily do this himself, he was covering the question: "And tell me officer, are you an expert in document examination?" To which he could reply: "No, but I know someone who is." I accepted this as a routine challenge, and all the Ts were crossed and Is dotted. The DC then turned to the accountant: "Can you follow the audit trail and demonstrate the nature of the fraud?" The initial reply was positive, and we were wrapping up the meeting when suddenly the accountant said: "Wait, it is not possible!"

We all stopped and the officer, looking worried, asked: "Why not?"

"Well," said the accountant, "it clearly says in the instructions that all seven copies should be made at the same time, therefore no one could put cardboard in between the sheets." We looked at him in amazement, and then Arne began to laugh. Our accountant friend looked hurt. It was no joke to him — he had been deadly serious. The DC took control with the memorable words: "This man is what we call a villain. He doesn't do what he is told or what he should." The accountant shook his head in disbelief and was probably cursing all scientists under his breath. Our report was submitted but, as is so often the case, we did not know the outcome.

Things that go bang in the night

In the 1970s the IRA was very active, as illustrated by the Birmingham bombings and the explosion on a train in March 1976 just outside Cannon Street. Not to make light of a serious situation, I was amused by an officer who contacted the laboratory to ask how to approach a suspect package "carefully", as instructed in a force circular. After much banter we decided that the best course of action would be to use a 'safety' match to illuminate the scene. This light-heartedness turned deadly serious when an officer who often used Scientific Services was called to a bomb at Baker Street station. He told me afterwards that he too had remembered the instruction to approach it carefully but was not in the least sure what that meant. We did have a couple of jobs involving suspected bombs, both of which, thankfully, turned out to be false alarms.

The sweepings from the floor of a mail van of the Irish Mail from Euston, which had suffered a fire, were delivered to the laboratory in several black plastic bin liners. The police were concerned that a distorted metal dial with partially melted plastic backing was found in the debris. Was this the remains of a bomb timing mechanism, the fuse of which had caused a fire, but which had failed to detonate? It took several painstaking hours to separate the debris to get an idea of what had been in the van.

There were no instant photos or videos in those days, and the debris had just been shovelled into the bags. Our examination revealed that in the vicinity of the distorted dial had been several small springs and a significant amount of thermoset plastic with charred cardboard embedded in it. It was only when spread out on the bench that it became apparent that what had been discovered was a set of kitchen scales, presumably packed in a cardboard box: the whole being wrapped up for posting. Why had it caught fire? The mail bags were manufactured from either jute or polypropylene. Jute will support smouldering which may eventually turn to flaming, and while polypropylene will not smoulder, but rather melt away from the heat source, if there is a flame it will burn vigorously. The most probable cause of the fire most likely a carelessly discarded cigarette, dropped during loading, onto a jute bag which developed into a flaming fire after several hours, which then spread rapidly when the polypropylene bags became involved.

The second suspected bomb occurred when a coach was burnt out at Ilford Depot. It being a time of heightened vigilance due to IRA activity, the bomb squad was called in. The area was flooded with officers, investigating the scene and carrying out house-to-house inquiries. They were concerned by the amount of (to them) inexplicable debris present in the coach, and it was suggested that scientists be called in to identify what it was. Why they did not ask the depot staff from the outset remains a mystery to me.

The bomb squad was probably thinking along the lines of their own laboratory staff, but a BTP officer suggested that railway staff may be more appropriate. When I arrived (by train, no cars then) I was confronted with a scene I had encountered on several previous instances of train fires: the skeletons of seats, damaged under-seat heaters and the remains of molten and distorted luggage racks. After much sorting, there was nothing to suggest any alien material was present and nothing to suggest an explosion rather than a fire. When satisfied that it was not a bomb, the depot staff examined the vehicle and rapidly concluded that the fire was due to an electrical fault. Better safe than sorry.

It's all in the cards

Credit cards were introduced in this country by Barclay Bank, using the brand name Barclay Card, in 1966. This was followed by Access, introduced by the Midland Bank, in 1972. Retailers including BR were slow to accept the concept, especially since in those early days every credit card transaction required a signature as verification and an image of the card details to be taken by placing the card on a bespoke metal tray, putting a carbon impregnated sheet of paper over the card and pulling a roller across it to imprint the card details onto the paper. These slips of paper were then sent to Barclay Card for payment to be transferred to the retailer (less, of course, the obligatory percentage paid to Barclay Card). To allow the tills at booking offices to be reconciled with the sales, a record of all Barclay Card transaction was kept, and the booking clerks were required to stamp BC on the back of the tickets, which were still the pasteboard Edmondson type.

The letters BC were contained within a circle which was also inked, and since the whole process was akin to using a John Bull outfit, it was easy to over-ink the stamp so that all was visible was a smudged inky mark with the pasteboard acting as blotting paper. Some clerks decided it was easier to stamp the front of the ticket rather than the back, and some noticed it was possible to obscure the ticket number and date of issue of the ticket if the BC was in just the right place. If the ticket could be returned to the booking office, it would be possible to re-issue it to a passenger paying cash and to pocket that cash, since the payment would be recorded as via Barclay Card and with a valid payment slip, but no money, in the till from its original purchase. Also, it would take several days and an enormous amount of paperwork to match each ticket with its appropriate Barclay Card counterfoil.

All that was needed was to get the ticket back to the booking office, and to sell a ticket stamped BC to a passenger paying cash. Both were achievable with a bit of guile and help from your friends. So successful was the process that the apparent decline in tickets issued at a certain London terminus station was noted by management, who could not explain why the actual number of passengers had not also fallen. One day, every queue at the booking office windows contained several police officers in plain clothes. If they received a ticket pre-stamped BC, they added it to a pile to be sent to the laboratory where both the ticket number and date of issue were determined using different light sources. Before the train had reached its destination, at least one ticket issuing window had lost its clerk.

A similar operation at another station involving catering store receipts resulted in the arrest of a clerk. Altogether some 400 documents had been falsely stamped as having passed through the accounts department while the money was passing through the clerk's pocket. He was prepared to plead to one offence: the receipt in his possession at the time of his arrest but would not accept the other 399-odd cases. Several days were spent verifying that all the receipts had been falsely stamped, and then several hours were occupied signing some 400 exhibit labels.

Copying is not allowed

Unbeknown to me, our work in document examination had been noted by the passenger department at the British Railway Board. I received a phone call from Mr Usher from the department, requesting I see him. When I arrived, I was told the board was very worried about the introduction of colour photocopiers. They had experienced several instances of people photocopying black-and-white first class season tickets; now they were concerned that the much more numerous green-coloured second class season tickets would also be copied. Mr Usher had arranged a meeting with Xerox and would I accompany him as his technical adviser. We are chemists, not electronic engineers, but it seemed unnecessary to mention this so I agreed and returned to the lab to read up as much as I could about photocopiers.

The day of the visit came, and we were made most welcome. During the conversation, while other BR representatives were occupied elsewhere, the Xerox technical guy, assuming I knew about all things photocopier, was extolling the virtues of their latest machine. He casually mentioned: "Of course we haven't got over the brown ink problem yet." I nodded sagely and asked what form the problem took in their particular case. "Well," came the reply, "it still appears black on the copy. But we are working on it."

He demonstrated the brown/black confusion and after the meeting, I wrote to the Passenger Department suggesting that all second class season tickets should have a thick, diagonal line printed across them in brown ink. The idea was accepted and for a short period, season tickets were so adorned until, true to their word, Xerox overcame their problem but in doing so, increased BR's. Does that count as forensic science? Arguably all it proves is that chemists are better at listening than at electronics.

The People's Palace ablaze

The London laboratory opened in 1960 on the site of the goods yard at the old Alexandra Palace station that closed in 1953 (not the current one, which is on a different route). This was the terminus of the line through Highgate from Finsbury Park, and the station was literally beside the palace with its own entrance to it. The lab was known as Fancutt's Folly since the Assistant Director of Chemical Services assumed railway departments would beat a path to its door, despite it being at the end of a residential road at a distance from the operational railway. They didn't: many may have got lost on the way (at least one police officer did), and a road vehicle was needed to go to Kings Cross every morning to pick up the internal BR mail, which was always addressed to Muswell Hill laboratory, c/o Kings Cross.

On the afternoon of 10th July 1980, Alexandra Palace spectacularly caught fire during the renovation of its world-famous organ, which involved more plumbing skills (with blow lamps) than musical aptitude. It took three days for the fire to be fully extinguished, but the most destructive time was the first few hours. I shall not forget the sight of corrugated iron roof panels, probably 8ft by 4ft, floating above the building, being carried on the thermals created by the fire.

The wind was blowing away from the laboratory but if it had changed direction, we would have been deluged with burning embers, not to mention the risk of decapitation from low flying roofing. The fire brigade was concerned that, as a laboratory, we must have lots of highly flammable, if not explosive, materials lying around and insisted that the building be evacuated. In other words, we all went home early. But there was concern about evidence in the laboratory.

Luckily most of it was suspect documents which were not very bulky, so I scooped the whole lot into two brown leather guards' bags. On my way home to Essex deposited the whole lot in the police office at Liverpool Street, after convincing the constable on the front desk that it was vital the bags were kept safe since they contained evidence pertaining to several important BTP jobs. The next morning, I collected the bags on my way to the laboratory which was now deemed out of danger of imminent explosion. Fortunately, there was no continuity audit otherwise we would have had to explain why evidence from all over the country spent an unscheduled overnight holiday in Liverpool Street.

Watermelons and a phone kiosk

It's no secret that in the run-up to the opening of the Channel Tunnel there was much concern about security, particularly regarding terrorist attacks. It must remain a secret (I have signed the act) as to what methods were to be deployed to secure the tunnel, precisely what it was secured against and from whom. However, by the early 1990s several firms, some of world renown and some completely unheard of, were touting their wares to the appropriate authorities. The lead departments involved were the Home Office and the Ministry of Defence, but British Rail was quite rightly involved since they would have to implement any system.

Part of that involvement included the BTP and the Director of Projects, a headquarters function that oversaw and/or facilitated major, typically one-off works. The Home Office and MoD have leading scientists and engineers in their employ, not to mention access to others throughout the world. The Director of Projects, not to be outdone, approached BR Research to establish what expertise we could bring to the table. The answer was not very much but try those chemists. Don't expect them to contribute anything very useful; in fact, don't expect them to contribute anything at all.

This was a very wise summation of our abilities in such matters. I was duly checked for my loyalties and let loose into a world where anything was possible financially, but not yet technically. While the real boffins slaved over complex solutions to identify and eliminate possible security risks, I tagged along to translate for the railway departments involved. And some of the solutions suggested really needed translation, mainly into the real world, from whence they had fled in flights of fancy of the security industry. However, some suggestions bore further investigation and even trials. This is where watermelons and a telephone kiosk come in.

If you want to detect a small but damaging amount of explosive in a large freight train and a firm in Silicon Valley says it can do it in a new way, then you'd no doubt wish to see what they have on offer. A firm in Silicon Valley said precisely this and claimed to be able to locate the merest trace of explosive, so we all — that is the Home Office, MoD, British Rail including BTP, and their French counterparts — went to see their proposed technique, flying out one day, staying overnight, conducting a test the next morning and then flying back in the evening.

We had the taxpayers' money to think about: two nights may have broken the bank, blow jet lag. But how to test it? If the technique was so innovative, there would be no standard test method. I don't know who came up with the idea, but it was brilliant in its simplicity: fill the floor of a warehouse with a cheap commodity, place the test explosive within one item, challenge the firm to find it and sit back to see what happens. The perfect cheap commodity in California was watermelon. But to make it more fun, do it as a double-blind test, meaning that the scrutinisers do not know where the explosive is located.

We arrived at the warehouse early in the morning to discover that the whole floor area was covered in watermelons, and the knowledge that one of them had been bored into and a small explosive charge inserted. The borers and inserters left the building so we, the observers, had no more clue of which melon was the target for the new technology than the representatives of the firm who were going to use it. The test began, the gubbins (the precise mechanism of which was still secret) was set to work and found... nothing. But the anonymous setters of the test had a final trick up their sleeve: the explosive could be detonated remotely by radio, and the detonator had been left in the hands of one of the senior observers. After a suitable time, to ensure that the technology had been given a fair crack of the whip, the radio signal was generated and a lone watermelon among hundreds of its comrades gave a violent shudder. It would be nice to say it exploded magnificently, showering the observers in gooey glory, but it was just a shudder. Use of the technique was not pursued, and watermelons may sleep in peace.

Now to telephone kiosks; well, a particular kiosk in rural Kent. An enterprising but very small overseas firm decided it had just the technology required for Channel Tunnel security. To prove it, they would rent a field in Kent and invite interested parties to witness their great discovery. This time the observers were only from the BR contingent. Recognising that the BTP would be able to identify where they had rented the field (Mark I eyeballs, trained to observe!), they were invited to drive straight there. However, these strange interlopers from Director of Projects and BR Research were a different kettle of fish. Could they be trusted with such top secret information as the location of a field? Probably not, so we poor souls were given the map reference (to ensure no confusion OS Landranger 2cm to 1km) of a telephone kiosk positioned on a triangle of grass apparently miles from anywhere.

We were told to be there at a given time, when we would be picked up. We duly arrived and at the time specified, a car drew up. What would happen next? We were invited to follow their car as it made its way to the field. This we did, driving for about 20 along narrow lanes and taking far too many turns; we were probably being driven around in circles to disorientate us. Eventually we arrived at the location of the test and invited to examine the demonstration they had set up prior to the trial commencing. We suggested a few alterations to ensure the test was not a fix and that the equipment had not been pre-programmed to work, and the test began. Either our suggestions blew the system or it did not work anyway, but the result was a dismal failure. We promised to write, if only we knew the address, but our verbal message was: "Thanks but no thanks." With far less intrigue we set off home, with no guide to follow this time, and I swear we passed the telephone kiosk a couple of minutes away from the field.

Vince's experiences may be unique in nature, but these one-offs were also part of the work Stinks were expected to carry out. Here Dave Smith shares his recollections of some of the cases the team dealt with at the Derby lab.

Just 50p to you

Sometimes we amazed ourselves at the speed we could respond to police matters. An officer from Birmingham BTP arrived with a pocketful of coins that were obviously fakes — they were in the unmistakable shape of the 50p but were completely blank. These forgeries were appearing in station slot machines and although the total recovered at that time was less than £100, the potential for problems was high. We asked him if he could wait and we entertained him for maybe half an hour or so while the analysis was carried out. The weight and chemical composition of the coins, which is unique to the Mint, indicated the blanks came from the Mint before they were pressed. The officer visited the Mint the same day and our suspicions were confirmed — someone was stealing the blanks. We never heard whether the thief was caught.

Another cheesy one

On one occasion Scientific Services were asked by Inter City on-board services to determine whether the coffee dregs in a cup were actually Maxpax, the brand name for coffee supplied by Maxwell House to the catering trade (it came in sealed paper cups pre-filled with a spoonful of coffee, to which boiling water was added at the point of sale), and whether a white deposit on a knife was cheese. We were concerned as to the cost effectiveness and the practicability of such a request, however, Inter City was most insistent. We contacted Maxwell House and asked them for help. We were amazed to hear they would be pleased to assist and support us.

The next problem was determining what constitutes cheese and how to distinguish it? This was resolved with the infamous statement that "the material had the consistency and composition that was consistent with it being cheese". Meanwhile Maxwell House confirmed that the coffee remnants were not Maxpax. But was it worth the cost? Armed with the information we had supplied, the on-board restaurant staff on the train from which the samples had been taken were convicted of fraud. They had been using discarded Maxpax cups and filling them with coffee from their own jar and making up their own sandwiches on the train rather than selling the railway stock. The outcome was amazing: there was £250,000 increase in revenue on the affected routes This indicated that fraud was being practised by more than one crew member. In fact, the demand for sandwiches was greater than the supplier could match!

We do drugs

As we moved into the 1990s and were able to take on external work, we ventured into drug testing for both the BTP and county forces. To undertake this, we required a Home Office licence to store restricted drugs on the premises. We were already fairly secure but we were now required to have safes fixed into the floor of the building to reduce the risk of ram-raiders stealing our samples which may have had significant street value. The visit by the Home Office inspector and the local constabulary put all sorts of requirements on us, far more than the police themselves observed; they agreed samples were often dumped on shelves within a police station. Returning to the lab one day by the back entrance, I was approached by a scruffy-looking woman with lank greasy hair and appearing to all intents and purposes like someone trying to gain entry. With some disdain, I asked if she needed help. She replied that she needed access to the building, so it was explained that there were security issues making this not too straightforward. She said: "I hope so!" and produced her warrant card. Her pockets were stuffed with samples from a drugs bust she had carried out in her disguise.

I ended up in a predicament when I went into the laboratory one Saturday to do a bit of work. With no one to remind me, I entered one of the security codes incorrectly and my entry had been logged at the local police station. I was happily working in my office when the phone rang. I was advised to come out with my hands up and was confronted by armed police stationed behind an upturned desk. I never went to work alone on Saturday again.

Dave's colleague at Derby, Roger Hughes, also has plenty to tell us about his experiences there.

Pull the other one

One of my early tasks was helping to curb an outbreak of communication cord pulling by football fans. Some fans were notorious and so were targeted by the BTP. Leicester was playing away, so I coated the back of the train's communication cords with malachite green mixed into sticky gel. We travelled down without incident, and the same on the way back.

At Leicester, I stood and watched as about 50 fans danced past the police, waving their hands in front of their faces like a Minstrel Show.

Following a successful job, four fans were brought to court in Manchester, accused of pulling the communication cord in a toilet. The prosecution case was presented and then the defence called the fans one by one. Each denied pulling the cord, saying they must have picked up the marker from the sink in the toilet. The prosecution lawyer targeted one of the defendants and got him to say that the four of them had been in the toilet at the same time. "If you were not there to pull the communication cord, why were four men in a public toilet at the same time?" he asked. He let this sink in and then gently began to speculate. The fan grew redder and redder as the court broke out into grins. Eventually the fan cracked and shouted that they had indeed gone in together to pull the communication cord, and for no other reason.

Beware of the dog

Behind the police office on Derby station used to be a pen and kennel, where an enormous police Alsatian called King was kept. Apparently, his handler's daughter was allergic to dog hair, so he had to live in the pen. I felt sorry for King and whenever I called by, I would take a biscuit for him, so he became quite friendly. Once I had carried out a fruitless communication cord job on a train bringing fans to Derby. As I walked beside the fans through the concourse at Derby station, there was a group of policemen and King with his handler, keeping the fans in a tight controlled group. I stopped to have a word. As King jumped up to lick my face, he trod on my foot and broke one of the small bones. It hurt and I screamed in pain. The fans, thinking a rabid police dog had got me by the throat, scattered wildly, throwing the police tactics into disarray.

Just a quick one

An officer I regularly worked with was DI Geoff Lawrence. He really dropped me in it at a Quarter Sessions, when he said there was no chance of being called to give evidence in the afternoon, so we may as well go for a pub lunch. A couple of pints later, we went back to court and the first witness to be called was me — Roger Hughes. Fortunately, it was a straightforward bit of evidence. Unfortunately, the recorder asked me how sure I was of my opinion. In vino stupidity I said I was positive. It was like watching a T-Rex getting ready to toy with its prey: he sort of unfolded his eyes, peered at me and said that expert witnesses usually couch their opinions in more measured terms. Could I expand? I floundered and looked over to Geoff for help, who was intently admiring the ceiling. I finally stuttered something along the lines of: "Would beyond reasonable doubt be acceptable?" Luckily the Recorder was prepared to accept that it would.

Geoff got his comeuppance in Corby. There had been a series of thefts of wagon bearings in the yards near the steel works. The thieves jacked up the wagons and lifted out the metal bearings, coming back time after time. The trap consisted of wiring together the back of the bearings so that if one was removed it would send a radio signal to the local police, who would raise the alarm. My job was to mark the bearings with a fluorescent varnish to provide positive identification if any were recovered. It was a big job involving the BTP and the Regional Crime Squad. Word came through that the alarm had been raised and thieves had made off across the fields with eight bearings, driving a Rover car.

The local police had a suspect, so about a dozen cops and I gathered outside a row of lock-up garages. "Which one is it?" asked Geoff. No one knew. Eventually a local ventured to say that the lock-up with the green door had a Rover inside. The lock was broken, the door opened and inside was a Rover, tight between shelves on the walls. Geoff ordered the boot be broken open — nothing. Break a window, release the handbrake and pull out to see if anything is up at the top of the garage, he ordered. This was in progress when an old man came along and asked what they were doing to his car. I've never seen a dozen burly policemen melt into the background so quickly, leaving Geoff spluttering and apologising to this by now incandescent Scotsman.

Back now to Vince Morris. As we have seen above, the culmination of many cases is an appearance in court so let's finish this chapter with a couple (well, three, maybe four — chemists can't count) of incidents in court.

Wet behind the ears

The defendants were charged with theft, using the simple expedient of printing self-adhesive copies of the labels used by a well-known mail order company, putting their own or a colleague's address as the delivery point and sticking the labels over the original ones. There is no need to dwell on how they got access to the original parcels. I gave evidence concerning a printing press found in one of the defendant's homes and remained in court to hear the rest of the case. When one defendant was asked why he ordered a car maintenance manual when he did not own a car, he told the court that he was working on a friend's car.

The barrister asked what was wrong with the car and the answer came back: "I was repairing a leak in the radiator." The barrister asked the clerk for the exhibit, and opening the manual read: "The Volkswagen Beetle has an air-cooled 1200cc engine." He stopped there, his face contorting in suppressed laughter. As it dawned on others in the court why he was laughing, others joined in — defence, prosecution, the public, the jury, even the other defendants. Eventually the judge cracked, and he too started to laugh. Only the defendant did not see the joke.

Walls have ears, courts have eyes

In a much more serious case involving kidnap, extortion, forgery and threat to murder, I was called as an expert witness to opine that travellers' cheques were signed by one of the defendants. Unusually I was not released (that is, allowed to leave) after giving my evidence, so was still in the court when a letter was discovered, the contents of which were significant to the defence case. The authorship of the letter was in doubt and the judge asked if I would compare the writing with that of the defendant suspected of writing it. It would, I felt, be churlish, if not downright foolhardy, to refuse an Old Bailey judge in his own court, so I agreed to act, as it were, for the defence. I needed control writing samples and the defendant's barrister arranged for me to go down to the cells to get the samples. I asked him if he was going to accompany me. "Not likely," he said, "he may nail my feet to the floor."

It was in the same case that the judge came to my rescue. I had given evidence about the travellers' cheques when, totally unexpectedly and I suspect irregularly, one of the defence barristers asked: "You are a scientist?" "Yes," I said. "If a gun was hidden between some bedding in a drawer, would you be able to tell it had been there?" Thinking on my feet, I replied, rather lamely: "That is beyond my expertise." The barrister, scenting blood, said: "Let me tell you about Locard's principle of transference" and proceeded to lecture me on the well-known idea that if A and B are in contact, traces of A will be found on B and vice versa.

When he finished, he said: "I ask you again. If a gun had been hidden between some bedding, would you be able to tell it had been there?" In some degree of panic, I replied: "That is still beyond my expertise." At this point the judge intervened, saying (and I paraphrase): "Members of the jury, the witness has quite rightly explained that this subject is not within his specialisation. You will ignore the counsel's question." One slightly worrying aspect of the trial was that I noticed an officer in the public gallery who had, I believed, no involvement in the case. Some time later we met on another job and I asked him, light-heartedly, if he had enjoyed my evidence. "Sorry," he said, "I wasn't listening. I was just there to see which villains in the public gallery were watching you." I'm glad I didn't know that at the time.

Justice never sleeps

Contrary to appearance, judges are not asleep. On one occasion when I was giving evidence about arson on a train, I said in passing that train seats are covered in wool-based fabric, which is naturally fire resistant. The judge looked up from his apparent slumber and asked: "Are you telling me that my wig will not spontaneously ignite?" "I am, your honour." "Thank you, that is very good to know," he said and returned to his somnolent pose. (I am, however, reliably informed that legal wigs are made from horsehair, not wool). A video we had taken to demonstrate the difficulty of accidentally igniting a railway seat was shown to the jury. It was the first time that video evidence had been used in that court and there was quite a crowd of clerks and ushers gathered to see how the process worked.

CHAPTER 5

Claims and Complaints

Valid claims or people making a point, we dealt with them all, from frozen potatoes and nuclear spillages to wild weed killing and things that go bump in the night. Dave Smith introduces the subject.

ANY LARGE public body will inevitably receive complaints about its services and claims for compensation, and BR was no different. It attracted some very bizarre letters and unreasonable claims that had to be dealt with in an appropriate manner. Of course, most people did not know of our existence, so their queries and comments were usually sent to the BR chairman or board in London. Unfortunately, they knew of our existence and passed the queries onto us with the brief diktat: "Please deal". In its early days BR was a 'common carrier' required to accept any goods, typically charging on a weight basis, and had many hundreds of thousands of goods wagons. Any goods damaged in transit would be duly inspected and if we were involved, a formal report would be written to support or reject the claim.

Complaints or enquiries from the public were treated with some caution, particularly when they were addressed to the Prime Minister, a Government minister, the chairman or BR board member. There was always a concern that the letter was not genuine, and that BR was being prodded for a response to be used in the Press to embarrass the organisation. A great deal of tact was needed but not always shown by our worthy chemists.

The following are but a few of the more amusing anecdotes of the many cases which each area lab could recount. They are based on the memories of Dave Smith in Derby (with the addition of a red-carpet contribution by Roger Hughes), and Vince Morris at London.

Over weed killing

One rather challenging environmental claim was made by householders stating BR staff had sprayed weed killer indiscriminately over their gardens. BR, as the railway still does, was required to keep the weeds down on the running track. It used a special train for this purpose that usually operated at night, spraying weed killer as it travelled along the specified route. A cocktail of materials was used in the 1960s and 1970s which would not be environmentally acceptable today. In stations and sidings, the train was not usually an option. Instead a gang of men with backpacks would be called upon to manually apply the weed killer. The complaint in question arose after a permanent way gang manually sprayed at one of the stations in the Birmingham area. Apparently, this gang was notoriously bad at any task: they had been given this job as it was thought to be one where they could do little damage. How wrong one can be!

To judge a complaint fairly, one required a poker face. On one hand your client was the railway manager who had summoned you, and on the other you were supposed to be arbiter. In fact, we were considered as very fair and independent, as evidenced in other anecdotes. I arrived at this particular scene and was escorted to the houses with the supposedly damaged plants, trees and crops. I met the owners of properties adjoining the track. The gardens were very long, and even the copper beeches which lined the road in front of the houses had defoliated in the middle of summer. It was if Agent Orange had been applied. The first observation was that the properties on one side of the station had complained — suggesting there may have been a strong wind in one direction (this was subsequently confirmed). Of course, the instructions were not to spray in such circumstances. In no way could I claim to be an expert on what weed killers could do, yet no one would have been in doubt as to what had happened. The damage was enormous and obviously not due to sodium chlorate, which was one of the main components of the mixture used on the weed killer trains. As I tried to keep my composure in the face of the evidence, the permanent way gang inspector said quietly in my ear: "It's bad isn't it? Shall I get the local manager to sanction buying anything they want in the local nursery?"

I took away a sample from one of the backpacks for analysis. In the laboratory, analysis showed that the weed killer — 2, 4, 5 T Brushwood Killer — was present in about 70 per cent concentration in the backpack.

As it was supposed to be substantially diluted, this concentration could presumably be devastating. I contacted Chipmans, the supplier, and asked what this concentration might do. Their technical man reckoned on a windy day it could kill tomato plants five miles away from where it was sprayed. I told the permanent way man to buy a few local nurseries rather than a few replacement plants.

Rolling out the red carpet

Roger Hughes recalls a simple solution to an oily problem. One claim involved a roll of red carpet scheduled to be used for the benefit of a Royal visiting Derby Town Hall, before Derby became a city in 1977. On the end of the carpet was an oil stain. My colleague and I spent the best part of a day and gallons of solvent trying without success to remove the stain. In desperation, we cut about two foot off the end and reported that we had removed the stain. I never knew if anyone at the town hall noticed the shorter carpet and trusted that it had not been made to measure.

Allergenic response

A lady wrote to a BR board member with responsibility for those with disability issues, stating she suffered from multiple allergy syndrome. She loved rail travel and asked that we advise her of all substances (vapour, dust or otherwise) she might encounter when she travelled by rail. The board member passed the request to the Chief Medical Officer (CMO), who rapidly passed it to us. This was an impossible task because, as well as railway contaminants from seats, brake dust, diesel fumes and so on, we could not be aware of what perfume or deodorant the person in the next seat might be wearing, predict the fumes from a farmer's burning field, what packet of nuts a fellow passenger might open and so forth. The only solution we could, tongue-in-cheek, suggest was for her to seal herself in a polythene bag for the journey. We passed this information to the CMO, who agreed with the conclusion and said: "I have no time for such people — I will send her your recommendation!" We never knew whether he did (knowing his character, it was quite possible), but maybe the business opportunity of providing sealed bags for travellers is something we should have followed up. I don't know what the response would be today.

I hope people don't think I am just a bag and put me in the luggage rack!"

Noise nuisance at New Street

Birmingham New Street station is enormous, and the extraction systems which dragged the air up from the lower level of the station where the trains run vented out at street level — not too far away from a high-rise block of flats. A complaint was made that the extractors were very noisy, and particularly noticeable in the still of night. Although there is some traffic noise at night, the levels that are deemed acceptable are substantially lower than those during daylight. There was a great deal of expertise in the field of noise within the research department's own Acoustics Unit, but they concentrated more on the theoretical aspects of noise from new or proposed lines and new rolling stock. We chemists in Scientific Services deemed that our competence, following training and experience in the field, was sufficient for this task. It turned out what we really needed was diplomatic and counselling skills.

"It's always happy hour here darling."

Phil Dimmick and I set up equipment at the location sometime after midnight to measure the noise output. All was going well until one or two residents from the flats came over to ask what we were doing. The numbers increased and we explained we were undertaking a noise survey on behalf of BR as a result of complaints from residents. They were not happy. They made it very clear they were not complaining about the noise and wished us to not make any recommendations to reduce it.

We asked what was going on, as the logic was strange. Apparently, some of the lower apartments were occupied by ladies who ran dubious business operations involving them attracting male clientele. They conducted price negotiations through the windows, shouting down to the pavement, and were having difficulty communicating with potential clients over the noise of the fans. The other residents did not take kindly to this trade and did not want us to make it easier for this nuisance business. Our report recorded the bigger picture and we never learnt what happened, but we also never had to deal with any more noise complaints from that source.

It went nuclear

Sitting at the desk dealing with fairly routine tasks was the norm, but sometimes we were faced with challenges we were ill-prepared for and we had to fly by the seat of our pants. We had no specific training on dealing with emergencies but when unforeseen situations arose, we were often turned to as the last resort. Such an instance occurred when the owner of a dismantling business in Leicester rang Scientific Services seeking help. He could have rung the police, fire brigade, Army or the Home Office but no — he picked us.

He had a problem. A railway parcel van sent for scrapping had a container within the vehicle with a radiation hazard warning sign on it. The container was yellow and about 40cm in diameter. The lid had become detached. There was a foam plastic insert for four vials within the open container. Two vials remained within the foam and were intact but the other two vials were on the floor of the van, with their contents spread around. After asking him to repeat his observations he asked if I was coming down. My immediate response was no but I told him I would send someone else! In the end, it was Peter Middleton who went.

The decisions to be made were many. If the situation as described was a real nuclear event then the site within the City of Leicester needed to be cordoned off by the police, various specialists called in and those who had come into contact with the van quarantined, cleaned up and checked for radiation exposure. Additionally, we should inform the Government's Nuclear Inspectorate and advise the BR press office appropriately to minimise the undoubted major panic which would arise. There were already concerns about the removal of asbestos from redundant railway vehicles carried out by the same contractor.

Sitting in my office, I decided that had there been a spillage such as the one described, there would have been alarm bells ringing before the parcels van was sent for scrapping. I contacted BR's Accident Investigation Unit and asked them to check up on the vehicle details and any incidents associated with it. I also contacted the rolling stock library at the headquarters of its owning department for information on where the vehicle came from and the reason for scrapping. It was explained that urgent answers were needed; a "I will come back to you tomorrow" response was not an option!

Peter reported that the container and vials appeared to be genuine but that the contents did not respond as if radioactive. Eventually all the inquiries I instigated came together. A training exercise involving the emergency services and centred around a supposed nuclear incident had been carried out in Wales. This included the simulated derailment of the van; the vehicle chosen was already destined for the scrapyard. When the exercise was over, the van was sent to Leicester for scrapping without clearing out the 'nuclear spillage', which was actually washing detergent. It had taken a lot of determined effort by numerous people to get to the truth. It would have been so much easier if the container had been removed at the scene, which is what I forcibly suggested to the gentleman from the Home Office responsible for the exercise, when he phoned a few days later asking for his container back. He was totally oblivious to the panic that his lack of foresight caused. The laboratory staff had given a very wide berth to Peter Middleton when he returned with his samples, just in case it wasn't detergent in the vials he had brought back after all.

Exhaust problems and solutions

One of the most frequent complaints to BR was about the environmental impacts of diesel locomotives. One such complainant was so prolific that he had a file dedicated to him alone. He was particularly focussed on emissions from the HST that formed the backbone of rail travel on several lines from the mid-1970s. The HST has two powerful diesel locomotives, one at either end of the train, allowing it to travel at up to 125mph. The writer was concerned that emissions from the locomotive were very deleterious to human health and decided that the best solution was to have two helicopters, each one operating directly above the power cars. Each helicopter would have a powerful vacuum collecting the 'toxic' emissions. He did not suggest there should be two more helicopters above them to extract the fumes they emitted, nor how we should capture the fumes when the trains were in tunnels! He did offer a solution though: his own patented device that could be fitted to diesel engines to ensure dangerous emissions did not occur. We wrote and thanked him for bringing the matter to our attention and filed his letter in his personal file.

Here Vince Morris recalls a similar complaint, and several others. A gentleman from Bath was concerned about the effect of HSTs on his city. He wrote to the chairman explaining that since the HST service was introduced, he had noted a marked decline in the moral standards of Bath's entire population. Since the only factor which had, in his view, changed was the presence of the new pointy-nosed trains, these were the cause. He went on to explain that he had written to the chief of police, the chief fire officer, the leader of Bath Council and his local MP but none had replied, so he thought the best course was to write directly to BR, starting at the top of the organisation. There is very little that can be said to such a person, especially when he had been so comprehensively snubbed by all his previous attempts to get authority to act but we felt it our duty to respond. Our reply was basically that he was talking rubbish but putting it that bluntly would probably solve nothing. We concocted a letter explaining that the exhaust from the HST was not much different from the exhaust of their predecessors and that he must be mistaken in his surmise that BR was degrading the country's morals. Either he was satisfied or he became so degenerate himself that he was beyond redemption; either way we did not hear from him again.

The rotten potato claim

Up to the 1960s, large volumes of seed potatoes were transported by train from the Scottish Lowlands to farms all over the country. They were often loaded into containers (pre-dating the ubiquitous metal ISO containers of today) which could be collected by the farmer at the receiving station and taken to the farm. The journey was often slow, with the final leg being one wagon 'tripped' to a siding and the station staff phoning the farmer to tell him to come and collect his spuds or pay for their storage. It was a favourite trick of farmers to complain that potatoes had been damaged in transit and for BR to receive a request for compensation (see photo 37 and 38). Often the area labs were requested by the Claims Department to confirm the legitimacy of the farmer's tale. At the London laboratory, John Salmons, himself a country boy, was known as Mr Potato because of the number of times he was asked to adjudicate between BR and a farmer. On one occasion he was asked to go to a remote farm in Kent to examine a load of potatoes which the farmer claimed had 'frosted' in transit and were therefore useless. The containers were usually packed with straw to provide some sort of insulation during such journeys.

John travelled to the nearest station and was collected by the complainant. He was driven deep into the countryside to examine the defunct tubers. It was soon apparent that the damaged potatoes were in the middle of the load, which suggested they were rotten when loaded, since if the frosting had occurred in transit it would be the outermost ones affected first. John made the mistake of telling the farmer there and then that he would be reporting to refuse compensation. The farmer suddenly became less than friendly and ordered John off his farm. John, of course, had no knowledge of local buses; indeed, no knowledge of which direction to go in to get to the station, any station. The next day, he regaled us in his rural burr with the story of how he was stranded in deepest Kent. He never did tell us how he got home, but from then on, he always told a complainant that his report would be "in the post".

It's not what you see

The word asbestos understandably strikes fear into the public psyche. When BR revealed the presence of the material in railway coaches as a sound and heat insulant (installed before the general concern about its use was raised and in locations not accessible to the public) there was justified concern. The safe removal of asbestos became a priority for the railways and various locations were selected to construct asbestos stripping houses. These facilities were approved by the appropriate authorities and meticulously managed and monitored, both within and in the

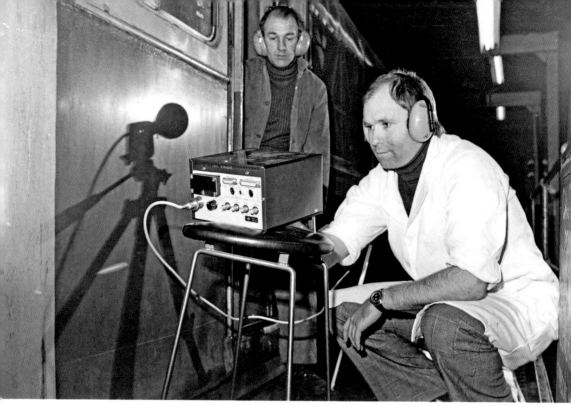

1: Not 'big brother' surveillance but measuring noise levels in a loco maintenance depot.

2: Checking to ensure that Sealink's Irish Ferries, from Stranrear, were not causing harmful pollution to mussel beds.

3: Sampling from Loch Ryan, with the boat skipper helping to look for evidence – of a Nessie?

4: Laboratory wet chemical work in the old days.

5: A classic piece of equipment in Glasgow Laboratory, the binocular microscope.

6: It looks like the chemist is checking an aquarium, but he is actually measuring oil viscosities as part of locomotive condition monitoring.

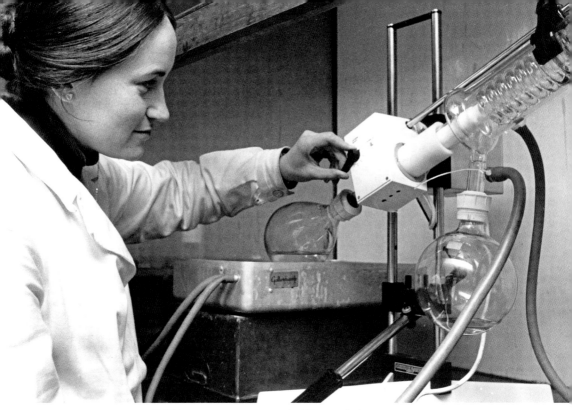

7: Using a rotary evaporator in Glasgow Laboratory.

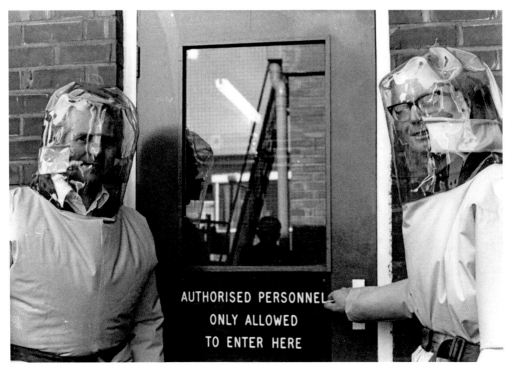

AUTHORISED PERSONNEL
ONLY ALLOWED
TO ENTER HERE

8: Scottish Region's General Manager and Chief Mechanical and Electrical
Engineer wearing air-fed protective suits, about to enter an asbestos stripping
facility that we monitored – not a Chernobyl nuclear site.

9: A photo of Derby Laboratory staff, taken not in an aircraft hangar but the first floor of Faraday House.

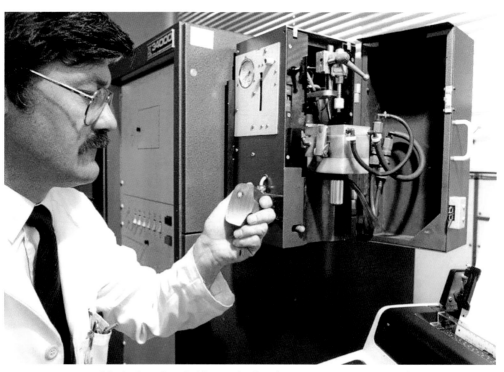

10: Not looking for a hidden stash of gin but operating a spectrograph.

11: Winding everyone up when operating an x-ray spectrometer...

12: The first infrared spectrometer used for the analysis of paints, plastics and so on.

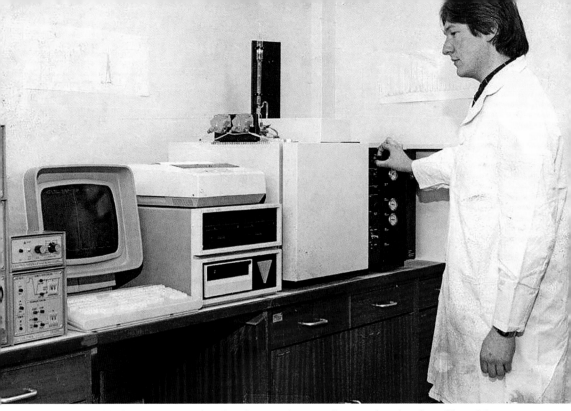

13: Modern instrumental analysis became the normal approach rather than old wet labs.

14: The laboratory in Calvert Street, Derby, circa 1950s.

REGIONAL LABORATORY (Sc)

ENVIRONMENTAL UNIT

PRODUCT & SERVICES UNIT

RADIATION PROTECTION DEMONSTRATION

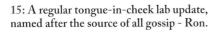

15: A regular tongue-in-cheek lab update, named after the source of all gossip - Ron.

16: The Doncaster Laboratory, showing oil samples, in the 1990s.

17: A rogue's gallery. In other words, the Scientific Services management team, circa 1990.

18: Kegworth Cutting in 1970, at the time of our early field tests.

19: Kegworth Cutting 50 years on. Very little has changed; the high walls and overhanging trees remain.

20: Two of the authors as young men, Ian McEwen and David Smith, sampling railhead contaminants.

21: A taped replica of a railhead showing a relatively clean central running band and rusty sides.

22: Plint 'portable' tribometer for adhesion measurement equipment by track teams.

23: Developing black rolled in leaf film on the head of a rail.

24: Bead of Sandite laid on the rail to improve adhesion on leaf-affected rail.

25: An attempt to produce leaf film for field tests.

26: It came from the sky; it was white; it looked like the regular stuff' – Daily Telegraph, 13th February 1991. The phrase 'wrong type of snow' was attributed to BR on this date, when they claimed that delays were due to snow that was dry and powdery rather than normal wet. Articles appeared in the popular press deriding BR with photos such as the one here from the Daily Telegraph.

27: The wrong sort of snow – dry powdery flakes that stick to surfaces and can penetrate air intakes.

28: A leaflet produced by Railtrack about leaves on the line.

29: Microscopic examination is a powerful tool for forensic investigation.

30: One of the authors, Vince Morris, in his younger days discussing a forensic case at London Lab.

31: Meeting with Scientific Services staff at the police training centre. Can you spot the Stinks?

32: The microscopic investigation, identification and matching of fibres, glass and materials.

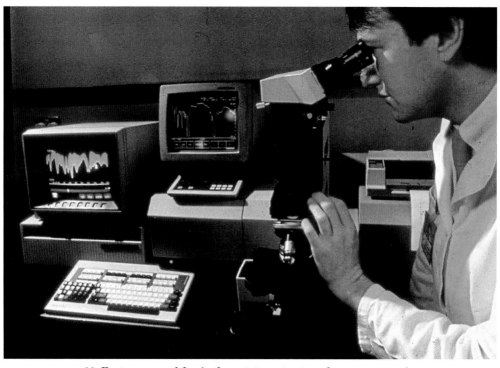

33: Equipment used for the forensic investigation of organic materials.

34: A promotional leaflet selling our forensic services for providing evidence in court.

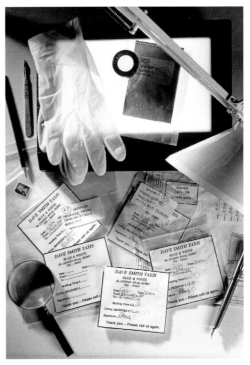

35: Promoting our services for fraud investigation of documents.

36: Arson investigation work.

37: Claims were made by farmers receiving apparently frosted seed potatoes from Scotland in the winter. Scientists were called in to inspect and sometimes arbitrate.

38: Potatoes, after frost damage.

39: An operative in personal protective clothing during decontamination work.

40: An instruction for passengers not to flush toilets when a train is stationary in a station.

41: Toilet waste discharged on the track at a station.

42: An Mk1 sleeping car. These were modified to use control emission toilets. Colour-Rail.

43: A drain for emptying controlled emission toilet tanks.

44: A dust cloud forms from a wagon dropping ballast onto the track.

45: An engineer walking through a dust cloud during ballast activities.

46: A burnt-out tamper at Pilmoor, Yorkshire, on 2nd April 1981.

47: A fractured hydraulic pipe.

48: Rapid spread of flames in Class 503 stock during a fire test at Derby.

49: A wooden crib constructed to simulate energy emitted by an electrical fault.

50: Once the fire started, there was a rapid spread of flames in seconds via the roof panels.

51: The coach's intumescent coating prevented fire engulfing it.

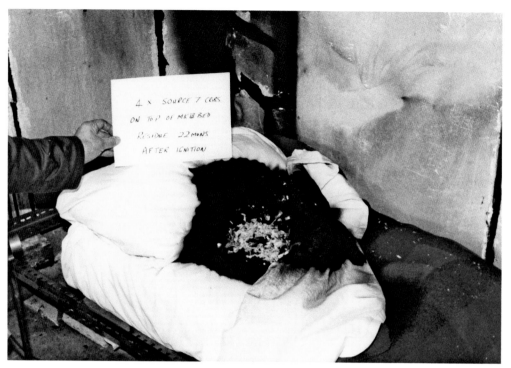

52: Fire testing of bedding for the Mk3 sleeper. A substantial ignition source failed to ignite bedding.

53: The fire on a Mk1 Sleeper in 1979 led to us setting up a fire technology unit.

54: Rail tankers catching fire within Summit Tunnel led to a
spectacular scene on the hills above. Mirrorpix.

55: The temperature within Summit Tunnel was such that it melted the tunnel lining. Mirrorpix.

56: Not feeding time for seahorses but measuring oil viscosities in the traditional way.

57: 'Spotting' oils for determination of dirt content. It caused Stinks to have spots before the eyes.

58: Measurement of spots by light transmission for dirt content.

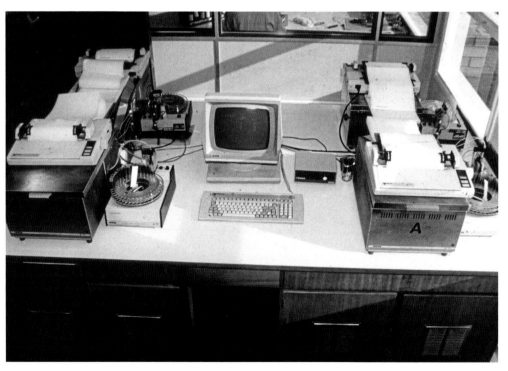

59: The early stages of automation for oil physical analysis to meet increasing demand.

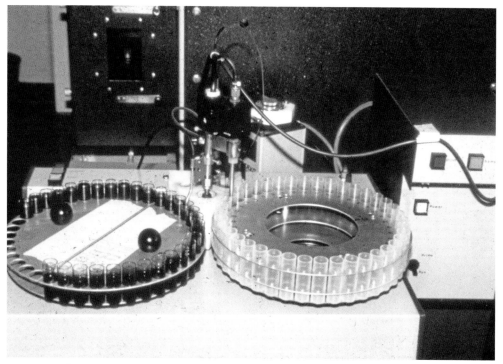

60: Sample carousels set up for automatic feed and spectrometric analysis of oil batches.

61: InterCity 125 HST reliability ensured by intensive oil analysis routines.

62: A close-up of the inside surface of the damaged piston boss,
showing the failed piston bearing (see below).

63: A damaged piston boss identified by oil
analysis before catastrophic engine failure.
When the gudgeon pin was withdrawn,
pieces of the bearing fell away.

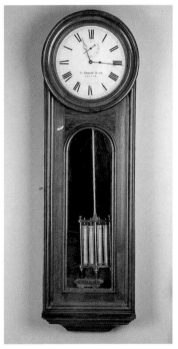

64: The chairman's clock was mercury
compensated; the vials of mercury can
be seen on the end of the pendulum.

65: It was nice of Thomas the Tank Engine (steam) to look after the fuel for the diesels! Checking fuel tanks was a regular job on depots.

66: Our cleaning technology team provided expertise on carriage washing plants. This one warns of the need to close windows and doors, otherwise a passenger might get wet.

67: A modern carriage washing machine in operation.

68: The pond at the Railway Technical Centre, where the unfortunate
gap year student took an unwelcome bath!

surrounding areas to ensure no fibres were released into the atmosphere. One such facility was established in the south of England, much to the concern of local residents. They hired an expert to fight on their behalf for the removal of the facility and he soon made his presence known to us in a professional exchange of correspondence. He was invited to view the operation, see our monitoring techniques, review our findings and take samples of his own. He accepted and became a regular visitor to the site. What he found was a safe, well-run operation with everything being done by the book and no potential adverse effects on the local population. But that was not what the citizens wanted to hear — they wanted the stripping house gone. The expert's report suggested he could not fault the procedures put in place by BR and he could not detect any asbestos escaping into the immediate vicinity. However, there was a suggestion that asbestos fibres too small to detect by any known means were the most harmful, and the fact that they had not been detected apparently demonstrated that asbestos was present. So, could BR please provide evidence that no asbestos had been detected and by so doing prove that it was there? Our answer was no.

The effects of asbestos inhalation is no laughing matter and to this day, the consequences of its widespread, uncontrolled use are apparent to all. BR took its responsibilities very seriously and extensive monitoring of the stripping facilities were undertaken. If asbestos was found in the atmosphere, the resulting clean-up operation was open, thorough and complete, with the public excluded from the area. But — and this is a big but — we did limit our monitoring, and subsequent actions to asbestos that was detectable.

The meaning of life

Maybe surprisingly, small quantities of radioactive materials can be purchased by the public (does your smoke alarm contain a radioactive source?) and were often delivered by rail. One such delivery was high-lighted by a letter to a freight manager. The correspondent complained that the material he bought had been damaged in transit and would BR please compensate him. Before forwarding the complaint to us, the freight manager made some initial enquiries and established that the package had been sent in the normal way, which involved a stop over-night while wagons were sorted for onward travel. The weather during the previous day had been warm and the wagon was unventilated. Armed with this information we contacted the complainant and asked him how the damage had manifested itself. The material was not functioning as it should, and he expressed the opinion that it might be due to excessive heat. We were intrigued. Normal temperatures, even those experienced in an unventilated van, should not significantly alter the properties of the material in question and certainly not its radioactivity. So, we wrote again and were again informed that the material was not working, and it must be BR's fault, because in every other respect the product was satisfactory. We wrote again and asked in what way the isotope was malfunctioning. To this day I do not know whether the reply we got was a genuine concern, or whether the writer had just given up arguing, because his reply was: "I bought the radioactive source to initially change lead to gold and then to discover the elixir of life. Since I have achieved neither aim, I can only conclude that BR damaged it in transit." We did not reply, and he did not continue his correspondence.

Baby it's cold inside

HSTs have been serviced at Bounds Green depot in North London ever since their introduction to the East Coast Main Line. Units awaiting entry to the depot were sometimes held at Ferme Park, just to the south of Bounds Green. Shortly after the introduction of HSTs, a resident of Ferme Park whose house backed onto the railway complained to his local council (Haringey) that he could not sleep at night because of noise from the idling trains. The council promised to investigate and as a courtesy asked Muswell Hill staff if they wished to attend the meeting. Haringey set up noise measurement equipment in the resident's house one night and everybody waited with bated breath for the readings. The resident suddenly announced he could hear the low frequency noise which had been keeping him awake. A quick look out of the window revealed there was no train in the area. The Haringey representative stood up, walked across the room, switched off the wall socket and asked: "Have you had that new freezer long?"

CHAPTER 6

Occupational Health and Safety

Even before health and safety assumed the importance it has today, the railways had been taking it seriously for many years. Here we look at spores, poo, fumes and more, with Geoff Hunt.

W HEN ROBERT Stephenson's locomotive won the Rainhill Trials in October 1829 with his entry, Rocket, reaching a top speed of 29mph, it was only a matter of time before the safety of this new mode of transport became an issue. It wasn't long before the first railway fatality involving other than construction workers occurred. On the opening day of the Liverpool and Manchester Railway in 1830, the MP for Liverpool, William Huskisson, was killed. Huskisson had reportedly been travelling on the same train as the Duke of Wellington but in a different coach. Their train, hauled by the locomotive North-umbria and driven by George Stephenson, stopped to take on water. Huskisson alighted from his coach and began walking along the adjacent track towards the Duke's coach. His intention was to ingratiate himself to the Duke, with whom he had earlier fallen out. He was unaware that Rocket, driven by the engineer Joseph Locke, was approaching. Rocket hit Huskisson, who subsequently died of his injuries. The 'new' railways were clearly dangerous places and in 1840 Her Majesty's Railway Inspectorate was established to oversee the safety of this revolutionary mode of transport and the industry surrounding it.

Safety became an important part of the industry's evolution, but it would take 130 years before the defining piece of UK legislation came into being — two years after the publication of Lord Robens's report on workplace health and safety. The Health and Safety at Work etc. Act 1974 introduced responsibilities for addressing the health, safety and welfare of workers across all industries. It also included requirements for organisations to ensure the safety of the public, which clearly applied to the railways. It was even later, during the 1970s, 1980s and 1990s, that concern for the physical health and well-being of workers as a result of their work activities led to the discipline of occupational hygiene evolving. The 'health' in health and safety was now addressed more formally.

Although safety in the railway industry on the operational infrastructure was well established with, for example, personal track safety training and codes of conduct and procedures for working on and around the track, occupational health assessment and management was a totally new field. This had a profound impact on the industry, resulting in Scientific Services being the obvious railway department to establish occupational hygiene expertise across their established Area Laboratory network. Key areas which featured in this early development — which applied equally to the railway workshops of British Rail Engineering Ltd (BREL) and the operations of the Director of Mechanical and Electrical Engineering (DM&EE), the Director of Civil Engineering (DCE) and the Director of Signalling and Telecommunications (DS&T), as well as train operators — included asbestos identification, monitoring and management (see photo 39); noise assessments and control; diesel fume emission monitoring; and dust and fume exposures controls from, for example, track working operations such as track renewal. All were carried out in compliance with emerging Health and Safety Commission guidelines and by the 1980s and early 1990s, these were further strengthened by a raft of European Union-led health and safety legislation, commonly known as the 'Six Pack'.

To meet this challenge, and after a number of high-profile accidents and incidents, in 1987 Scientific Services established the Derby-based Health and Safety Unit, which brought together specialists in the various health, safety and environmental management disciplines a 20th Century railway business needed to address.

The unit ultimately comprised of occupational hygiene specialists, fire technologists, human factors researchers, derailment investigation engineers and members of the traffic and dangerous goods team alongside emerging environmental management specialists. Its remit was wide, including such topics as providing ergonomic design advice for the next generation of railway control rooms that replaced the signal box, ensuring that materials used in building rolling stock for use (e.g. in the Channel Tunnel) met stringent fire safety standards and making sure that loading and unloading vehicles for the Eurotunnel shuttle service did not exceed exhaust fume standards. It was also part of the remit to ensure that employees' awareness of occupational health risks in the workplace were addressed through the provision of appropriate information, instruction and training and, where additional control measures were needed, ensuring the provision of personal protective equipment or engineering controls were fit for purpose. The unit's environmental management expertise addressed emerging problems of contaminated land management, and the decontamination and monitoring of polluted land, air and water course emissions. It advised on how to protect the existing railway infrastructure, prevent pollution from new railway developments and the protection of the existing natural environment (flora and fauna) from current and proposed railway operations.

As the railway industry entered the private sector, these core support services continued to be provided by Scientific Services. Railway industry occupational health, safety and environmental management control had come a long way since the fatality of the unfortunate MP for Liverpool.

LEFT: *"You are being transferred because the asbes-tos ceiling tiles could affect your health."*
RIGHT: *"Worried about my health? I'm more worried about the safety of losing my head to an axe tomorrow!"*

Anthrax, more anthrax and the Black Death

It started off as a simple task for a railway buildings inspection engineer: just go into loft bay at King's Cross station, check for leaks and damage, do your report and the job's done. Imagine the scene as the inspector carries out his routine risk assessment…

"Access to the loft hatch is good, will need additional lighting but have a torch for the inspection and there are good walkways around the loft bay to get where I need to go. There is some insulation material too, looks like fibre glass but better check and send a sample to Scientific Services. They supply their normal good service and report back quickly. The report says the insulation is fibre glass and horsehair. Well, didn't expect horsehair, never seen that included in the loft bay insulation report before, so had better look it up for any additional risk while the inspection is undertaken. The HSE is the place to go to find out about these odd identified hazards, so just what harm could horsehair actually do? The answer: anthrax. Definitely back to Scientific Services now, who can turn their hand to anything."

It was quickly established that horsehair was a common building material in the 19th and earlier centuries, used to reinforce material for 'plaster and lath'. If you look closely at the walls of old buildings, you can sometimes see the individual animal hairs in the plaster mix. But as the buildings at Kings Cross are more than 100 years old, for anthrax still to be present after that length of time is surely most unlikely. However, recognising our limitations in this area we sent samples to a laboratory that provided an analytical service for identifying anthrax, the Health Protection Agency (HPA), located at the secure Ministry of Defence facility at Porton Down.

The report Scientific Services received from the HPA was definitely not what we expected. Yes, there were anthrax spores present, old dead ones but also within the fibre glass were small numbers of viable anthrax spores — 'anthrax daughters' as the HPA described them. So what risk do they actually present? Again, we turned to the HPA for a more specialist technique of risk assessment for this potentially dangerous material. Their recommendations were clear: the risk was small but still potentially present, particularly with respect to inhalation of the anthrax spores when people were moving in and around the loft bay while undertaking inspections and any subsequent repair works.

This was a new area for Scientific Services but not insurmountable as it had long experience with asbestos management and its control, and this was a similar scenario. Liaising with building inspection engineers and the station owners, at that time Network South East, Scientific Services drew up a safe system of work for inspections. The risk was small. Although anthrax can enter the body via the skin, or through ingestion or inhalation, it was only the latter that required controls and as an additional precaution, to prevent entry through the skin, any open wounds were to be bound with sticking plaster. The system was quickly established, based on those used for working with asbestos and employing appropriate control measures to handle any potentially anthrax-contaminated personal protective clothing. Using our guidance, the loft bay inspection was carried out and repairs identified. This presented the building engineers and owners with more problems. Back to Scientific Services — only this time the HM Railway Inspectorate became interested and wanted to know how British Rail was going to manage the potential risk.

Following numerous meetings and discussions, including threats of enforcement action by the inspectorate, a procedure was established between Scientific Services, the building engineers, the owners and the Railway Inspectorate for the removal of anthrax-contaminated insulation and decontamination of this particular loft bay. The area was sealed as tightly as possible under negative pressure, with air being extracted through highly efficient particulate filters. Workers were protected against inhalation exposure or skin contact and, finally, all the surface areas were wiped with a dilute solution of sodium hypochlorite (bleach). Air samples were taken and analysed for anthrax spores by the HPA before the loft bay was reinsulated with fibre glass and returned to normal access arrangements. The potentially contaminated insulation and used personal protective clothing was double bagged and removed to a specialist incineration contractor for disposal above 1000°C. Consideration was then given to the remainder of the building's loft bays. Fortunately, this initial removal exercise went well. While some viable anthrax spores were initially found at the inspection stage, subsequent analysis reports indicated that none were present in either the swabbing surfaces or in the air extraction filters. At a review of these findings with the HPA and the inspectorate, it was agreed that future works would only require the normal personal protection used when working with, or in the presence of, glass fibre.

This was not the only time anthrax was flagged up as a potential problem in the industry. In the 1980s, the company Union Railways was established by the Government with the aim of building and operating the new Eurostar hub at St. Pancras (before it arrived years later at Waterloo). Scientific Services became involved with the initial site surveys to determine what potential health hazards should be considered before the site was cleared and construction began. The initial work involved carrying out a historical desktop survey and site sampling before physical work commenced.

This survey revealed that St. Pancras and the surrounding area had long been an established industrial area of London dating back centuries — and the investigation identified a potential risk for anthrax to be present. This was because the site had once housed a 19th century horse infirmary which served the local industry and canal network operators, and there was the potential for anthrax to be present from the veterinary activities and from animal hair and skins that may have been present on the ground or buried on the site. The desktop exercise also revealed that the site had been a 14th century burial ground from the period of the Black Death! Could there be any remains still present and could the bacterium responsible for the Black Death still be there? The Union Railways project never went ahead at St. Pancras, so no further technical or analytical work for confirming the anthrax and Black Death contamination were needed. Could there have been a risk from residual Black Death bacteria? Well maybe not, as it was thought that in the case of this burial ground, the coffins would have been sealed and lead lined. But who knows?

Loos and poos

An unsavoury topic, but someone has to do it. Here, Ian Cotter introduces the subject.

From when train toilets were first installed in carriages, they usually discharged onto the track until newer trains built in the last quarter of the 20th century were fitted with a retention system. Toilet discharge could cause problems for track workers, either as the train passed or when they came to inspect or repair the track. It was also unpleasant for depot staff whose job it was to clean the undercarriage prior to maintenance work being undertaken. There was the added issue that toilet emissions also tended to spray the side of the train with the detritus when the toilet was flushed. There was the chance that a passenger may lean out of the window to unlock the door (most train doors could not be opened from the inside) only to find a piece of soiled toilet paper and/or other debris wrapped around the handle! There were complaints from time to time, but the then British Railways Board's Chief Medical Officer, not very helpfully, pronounced you would need to eat a walnut-sized piece of faeces before it would cause a health problem, and he concluded that this was unlikely to happen.

There was sometimes offensive-looking debris deposited on the track by those ignoring the 'Do Not Flush In Stations' signs (see photos 40 and 41), sitting on the track in full view of waiting passengers. Consequently, a piece of equipment was developed by the Doncaster laboratory that used a pressure washer with a hood on it that platform staff used to encourage the debris to wash into the ballast out of sight. Some stations like Paddington were designed with brick troughs and staff there could be seen in the middle of the day hosing down debris from sleeping cars deposited the night before. Accompanying this work was a range of tests carried out to establish what residues could remain after a toilet flushed onto the track. Staff from the London laboratory travelled up and down on local services, flushing and then swabbing door handles for bacteria as the journey progressed. How the offending material was obtained to seed the toilet waste is unknown; maybe the selected technician was given a vindaloo curry on expenses beforehand. It's more likely that bacteria from the swabs were grown in the microbiology laboratory to provide some reproducibility in the tests. The conclusion was that above about 20mph, the handles were effectively scoured clean by the air stream unless there was a trapped wad of toilet paper or other solid waste present. The project then progressed to examine a system whereby toilets stored waste and then discharged it when the train's speed reached above 20mph. Sleeping cars were seen as the biggest problem and one was modified to retain all the toilet waste using the Elsan system beloved by caravanners and campers, which was then released at depots into their drainage system. This was the forerunner of the controlled emission toilet on modern trains, which have bigger tanks and are discharged only in depots specially equipped to deal with them.

Here Roger Hughes tells of his experience with one Elsan type toilet exercise, potentially for use in the advanced passenger train (APT). Since such toilets were mandatory on boats on the Norfolk Broads, one was modified by BR Research's Plastics Development Unit (PDU) to incorporate a clear Perspex bowl, so its effectiveness could be observed. The bowl was bolted to the waste inlet and outlet and a macerator was installed that chopped everything up. The chemist was now ready to test the modified unit. The night before the test, in preparation, a curry and a few pints were partaken, all of course in the interest of engineering and science. The next day the modified bowl was loaded with the fruits of the previous night's activities, bolted in place and switched on. The result was horrible. My mistake was that I had relied totally on the PDU but they had cut corners, replacing the original 18 stainless steel bolts and thick gasket with 12 nylon bolts and a thin gasket. Under pressure, the unit acted like a dozen high pressure water pistols and unfortunately, clean tap water was not being used. Despite repeated showers and changes of clothes, this poor chemist was ostracised for weeks.

Dave Smith takes up the story now, explaining the development of the controlled emission toilet. Apparently, an order was placed on BR by Barbara Castle when she was a government minister in 1969 to stop the fouling of the track at King's Cross with toilet waste. Although it sounds horrendous, the evidence was that the waste became harmless in a very short while due to light exposure and did not pose a serious risk to health. However, as those who work on the track will tell you, it is not a pleasant experience when a train passes and the toilet waste baptises you for the first time — or at any time, for that matter. Likewise, when a train passed through a station, flushing a toilet would be unpleasant for a passenger on the platform in the wrong place at the wrong time.

The problem in Kings Cross was caused by the sleeper car trains which left late in the evening to Scotland. There were up to three trains in the 1970s leaving somewhere between 22.30 and midnight. Passengers were allowed to join the train early and make themselves comfortable. As well as having a few drinks and snacks at the bar, this usually meant a trip to one end of the carriage to two toilet cubicles. There were only two toilets for typically 22 passengers per sleeping car and were often used contrary to the instruction to not flush in the station (see photo 40 and 42). Plying people with food and drink and not expecting them to use the toilet seemed a strange combination. There was a chamber pot for liquid waste in each berth hidden from view under the sink; all right for men but not so convenient for women. They were not used much, hence detritus from the toilets formed very unsightly pyramids of waste on the track which, apart from being visually unattractive, was considered a major health hazard within the confines of the station. The trains also tended to be positioned in about the same spots each night, causing the pyramids of waste to grow. It was certainly unpleasant for those who serviced the vehicles and it was unacceptable to the public during daylight, when the problem was clearly visible.

In an effort to resolve this, it was decided by a collaborative European rolling stock committee of which BR was a member that the vehicles of concern should be fitted with a control emission toilet, or CET, collecting toilet waste in tanks beneath the coach. A small implementation team was established which included design engineers, on-board operating staff and depot maintenance staff. This was in the mid-1970s when the Health and Safety at Work etc. Act 1974 was introduced, and it was therefore thought that Scientific Services should join the team to advise on the potential occupational health and safety implications for such a sensitive issue — particularly for engineering staff potentially coming into contact with gallons of human waste and excrement during the emptying or maintenance of the units.

The design engineers were told of the need to modify the vehicles. However, the idea of consulting and working with others to estab-lish whether their design was sensible or practicable was not common practice. Wonderful designs were created that were impossible to either build or operate. The implementation team met when production was in full swing at the BR Engineering Limited (BREL) works in Glasgow, making the various components and modifying the vehicles accordingly. Many vehicles were being taken out of service and modified by the time the first meeting took place. One standard coach was modified for trial purposes and available, if needed, for tests. Modification involved collecting effluent from each toilet into a small tank and when full, transferring it to a larger tank slung under the middle of the coach. This tank could then be emptied as required, but there was no indication as to how the emptying should be achieved. This meeting was supposed to ensure that any operational problems were identified and eliminated before the modified vehicles went back into service.

Basic questions were asked of the engineers:
1. How much water was used when you flush the toilet? Answer: One gallon.
2. How many people were on a vehicle? Twenty-two berths were available.
3. What were the volumes of the small and large tanks? Five and 20 gallons.
4. How often would it be expected that someone would use the toilet? We do not expect people to go more than once! After a drinking session, we asked, and it was agreed that the large tank should be increased to 50 gallons.

The idea seems simple enough, but the detail showed a thorough lack of foresight. For example, how were the small tanks and associated macerators to be serviced should they fail or need to be maintained? The answer was, unbelievably, that they had fitted a trapdoor at the bottom of the tank which could be opened by a maintenance fitter if he went underneath! Someone did eventually ask me if this was satisfactory.

It was apparent that such an arrangement was totally unacceptable and that trade unions, flexing their muscles and quite rightly using health and safety as a reason, would make sure the railways came to a halt if such an arrangement was ever introduced. The point was made that, even as the committee considered the problem, BREL was making and fitting these tanks. It was irreversible and if there was a problem, what was to be done about it...? Ask Scientific Services. This was the start of an interesting exercise involving colleagues from other railway departments. The mistake was made of asking what facilities were available at depots and workshops to operate the discharge or maintain the system. This was obviously another silly question. The answer was: nothing.

Engineers went away to work on a modification. After a suitable interval, a visit was arranged with a rolling stock design engineer to examine their modification at Bounds Green Depot in London, where a modified vehicle could be demonstrated. Off we went, the implementation team smartly dressed for the occasion and with a party of BREL and BR dignitaries to review progress. The modified vehicle had been in operational service and so this was considered a fair trial. The engineer was confident: the small tank now had a side door that could be accessed without going beneath the vehicle. He was so confident that he was happy to personally demonstrate the equipment. He leaned forward and released the trap door mechanism, and the contents of the tank — liquids and solids — shot out with some velocity. Unfortunately, a significant amount of it went up the sleeve, inside and outside, of his smart suit. To say the smell was not very pleasant is an understatement. We travelled back separately to Derby.

When trains operated, there was a sleeping car attendant present for every two coaches. It was their job to swing a lever that emptied the large tank when the train was in motion within a short while of departing Kings Cross, leaving the tanks empty for subsequent night-time use. Did this mean that tanks of effluvium (300 gallons per train) might be discharged on a regular basis at Alexandra Palace at about 10.30pm, followed presumably by two more trains with a similar pay load? Why should this solve the problem, since it was still discharging the debris onto the track? It was true to say that toilets had typically been designed for track discharge. However, discharging 300 gallons of diluted human waste and excrement some three times a night in roughly the same place was not quite the same as the odd flush. Objections would inevitably be raised by anyone standing on platforms north of Kings Cross at night when this was done, just as the event from the odd flush from a passing train was never too pleasant to a passenger standing too close on the platform.

"Oh shit!"

Inevitably there was some difficulty in convincing the implementation team that this was unacceptable. To demonstrate the point, it was decided to run tests using one such modified vehicle filled with whitewash in the tank. The next task was to demonstrate the situation to the satisfaction of all. A straight piece of track in the Yorkshire countryside was located, where the tank would be discharged at speed and filmed using a high-speed camera. Scientific Services personnel would be at Kings Cross, ready to take further pictures of the train when it arrived, to demonstrate the impact.

The occasion was memorable! It was a fine day and an ideal temperature; a wet day may have required us to do the test again. There were fields around us and a herd of black cows. Surprisingly, everything went to plan. The train was on time, our staff were on board ready to execute the manoeuvre with the whitewash mixed in with contents of the tank, the communication systems worked and so it was all systems go. The camera was set up some distance from the track and staff were there alongside the equipment, ensuring we got the best view and hopefully did not miss the outpourings.

In fact, there was no prospect of missing the outpourings. It was visible from some distance as the discharging train approached us. The cloud was so big that we ran as far away as possible… and black cows became white! What the farmer thought when he rounded his herd up that night is anyone's guess. He probably had a few drinks on the story about the aliens whitewashing his cows. Since the discharge vehicle had not been marshalled to be at the rear of the train, staff were equally astounded by the new livery colours of the train arriving at Kings Cross.

The next meeting of the implementation team was very quiet. "Where do we go now?" was the question around the table. The result of this debacle was a complete re-think. Clearly emptying the tanks while the train was moving was a non-starter; it had to be done in a depot. The two main BR depots on the east coast route had concrete aprons fitted, at enormous expense, where vehicles could be emptied, and waste drained away to the foul sewer.

There were still problems to be addressed. How could we ensure the design functioned satisfactorily and how could the CET be cleaned and maintained appropriately? These questions were tackled with the usual Scientific Services verve: project evaluation undertaken by a lucky chemist — Geoff Hunt. Initially, a large tank was provided and connected to the mains water supply at the Derby Hartley House laboratory. Simulated human waste options were fed into the tank and then it was emptied to assess how effective or otherwise this would be in a depot (see photo 43). Would the valve block? Would the waste drain? Sawdust, whitewash solution and various papier mache mixtures were used to simulate human waste, but it was soon apparent that these could not replicate the real thing.

As a result of this initial trial it was agreed there would also need to be significant investment in steam cleaning plants in the various railway works and depots before the vehicles were subject to maintenance and repair. Steam cleaning was chosen as a safer and cheaper alternative to the use of chemicals used in the earlier Elsan exercise. As part of developing operational methods for the depots, we were also actively involved in establishing the time needed to kill the expected bacteria with steam jets and the working temperatures required associated with the CET installations.

Consequently, an actual tank from a coach was fitted to the outflow of the Hartley House laboratory gentlemen's toilets, which was connected to a bowser to empty the tank to the foul sewer. The tank was also fitted with a viewing panel and internal windscreen wiper on top to assess the constituency of the contents and to indicate when it needed emptying. The test soon confirmed that the human digestive system does not readily digest tomato seed and skin! Soon the male occupants of Hartley House objected, claiming it was an intrusion of their privacy, and avoided the cubicle whenever they could. This was overcome by always inviting male guests and visitors to the building to use the facilities and thanking them for personally participating in one of our scientific experiments. Some were not amused! Eventually, to get sufficient liquid and solid matter it was necessary to connect all of the cubicles and the urinals to the system.

A further modification was the addition of a macerator to break down the larger deposits of the contents, taking account of all the other nasties people might chose to throw down the pan, so that the emptying process would not be delayed by blockages. We also determined which type of valve mechanism would achieve maximum unrestricted output during the emptying process, settling on a two-inch butterfly valve design.

There is, however, only so far that simulations can mimic real life. There were still the synthetic contents a toilet would have to contend with. You can imagine the face on our Procurement Department when we raised a regular order at Boots the Chemist in Derby to stock up with a gross each of sanitary towels, nappy liners, disposable nappies and lots of condoms. Submitting a petty cash claim to the Finance Department without any explanation was obviously going to cause ructions but internal procedures had to be followed and could be challenging. You needed to give them some excitement in the offices from time to time! The resultant telephone conversation was priceless. "What the hell do you think you are doing?" or words to that effect. "We are not here to supply your household with all its toiletry needs, never mind condoms." The response was similar when a male railway chemist turned up at the Boots counter with a regular bulk order request: "Really? For British Rail Research?"

The project was now nearing completion. The equipment was operational and for some six months, regular feeding and emptying of the system, reporting interim findings and hosting displays to interested parties, fell on the shoulders of the lucky Scientific Services chemist. The final task was determining the decontamination process, which had been previously agreed as steam cleaning because chemical treatment was considered much too hazardous and potentially expensive. Cleaning was required before a tank could be worked on without placing engineers at risk from the contents, particularly as there were now increasing concerns about contamination from more dangerous pathogens such as the Hepatitis family and HIV.

Determining when the tank was safe to work on would be undertaken in two parts: heating the soiled empty tank using a steam lance while monitoring the external metal surface temperature, followed by taking discharge sample swabs at 15-minute intervals from the waste outlet for subsequent microbiological analysis. Our chemist was suitably attired in a bright green protective rubber suit, face visor and gloves, similar to what depot staff would wear. A suitable day was chosen and decontamination commenced, running for some six hours until the tank surface temperature reached 70°C, which our microbiologist specialist said could reliably be expected to eliminate the biological risk (to be confirmed by analysis for E. coli in the discharge samples). This enabled a graphical representation of the reduction of risk as the bacteria count diminished. The odours coming off the tank at this time were ripe, and as the tank was connected to the laboratory toilets the chemist was not very popular with the rest of the building's occupants.

The analysis completed, a report was compiled on the 'Use, Operation and Decontamination of Control Emission Toilets' by our lucky chemist. The outcome was successful, and a full programme of installation commenced. The project had come a long way from the initial exercise at Bounds Green depot.

By royal appointment

Here, Dave Smith explains why he is still waiting for a letter from the Queen.

Several years after our work on CETs, I received a nice phone call from a very well-spoken gentleman who advised that he was the captain of Her Majesty's royal train. He said he had a delicate matter to discuss and it concerned the toilets that had been modified on the train. Apparently, they were working well apart from one problem: the discharge and smell from them was not nice, and what would we recommend? Our chemist got to work and quickly found a royal blue coloured deodorant (or re-odorant) for adding to the tanks. The 'By Royal Appointment' sticker never did arrive.

Ballast dust

It has long been recognised that the permanent way staff working on track activities could face potential health risks. These come about in various ways, from the risks associated with moving trains and heavy equipment to those arising from the engineering work itself, e.g. noise or dust emissions and other contaminants from ballast and the track bed.

In 1978 Scientific Services was requested to undertake a survey of dust produced by ballasting operations (see photos 44 and 45). The railway track consists of steel rails supported by concrete or wooden sleepers on top of a bed of ballast. Dust could arise from track machines operating on the ballast to improve the track, the dropping of new ballast from wagons or ballast cleaning activities, where a machine removed the ballast and returned the good ballast to the track. If dangerous concentrations of dust can arise in the open, consider what could happen within the confines of a tunnel.

The ballast obtained from quarries around the UK at this time was typically limestone, basalt, gritstone or granite. From a risk-to-health perspective, dust is considered a nuisance in its own right. It can be a health hazard, particularly if it is very fine (not visible to the naked eye) and it contains quartz. Working environment exposure limits were set (and still are) by the Health and Safety Commission (HSC) as to what was an acceptable risk. The limit for what was defined as 'nuisance' dust then was $10mg/m^3$ (milligrams per cubic metre) and the limit set for respirable dust was as low $0.1mg/m^3$. Strangely, the limit set in general was much less than that for coal mining operations, where personal exposures could be much higher. That figure was $0.45mg/m^3$.

Scientific Services used its laboratory resources around the country to investigate multiple railway ballasting operations using different types of ballast. This involved airborne sampling of various track worker activities and taking actual samples of the ballast for chemical analysis. Granite typically had a significant alpha quartz content yet the respirable content when ballasting was remarkably low in comparison with the limestone ballast, which had very low quartz content but gave rise to high dust levels during operations.

Working on the track is not without risk in itself as it is uneven, and many rail maintenance activities are carried out at night to avoid service disruption. The dust could obviously be a visibility issue, but the focus of this survey was just the inhalation of airborne dust. Wagons dropping the ballast sometimes became derailed if too much stone was dropped, and once one of the chemists carrying out the sampling exercise was thrown off a vehicle. So, this activity was recognised as one that posed significant risk.

The survey results showed that very occasionally the dust levels could be unacceptably high, often only for a matter of minutes but still exceeding prescribed HSC limits. As a result, it was recommended that all trackside equipment used in ballasting should be fitted with filtered pressurised air so levels within vehicle cabs were well below the specified HSC limits. In addition, vehicles and persons who did not need to be in the area were to be excluded.

Engineers working on the track in the open air were recommended to wear a simple ori-nasal face mask when it was very dusty, usually for a period of less than 20 minutes. This was before the days when the use of personal protective equipment was controlled by its own individual piece of legislation. Although this had the advantage of minimising overall ballast dust exposure, it would only prevent the inhalation of larger particles of dust. In theory, the only acceptable protection for the high levels of dust occurring very occasionally would require workers to wear full self-contained breathing equipment, such as that worn by firemen, and would necessitate changing cylinders every half hour. This was viewed as impractical and posed its own safety risks to those working on or walking the track. The Chief Medical Officer was subsequently asked to provide information for track workers and evidence of dust problems associated with quartz exposure. We were advised that there had been no cases reported and hence the provision of full breathing apparatus could not be justified.

What the report did not include was the observation that many of the workers reported for duty straight from the pub on a Saturday night at 22.00 for a 12-hour shift, and dust was not the greatest risk in our opinion. This was long before the days of the Transport and Works Act 1992, which stopped this out-of-work activity!

There are countless stories of derring-do associated with dust and fume monitoring. The following two tales here are from Geoff Hunt and the latter two from Vince Morris.

All industries have their eccentrics and Scientific Services was no exception. In the Hartley House upstairs laboratory, it was customary for breaks to be taken in room 109, where meetings were held and where staff without an office had a desk for writing reports. During one such break period, the door burst open with our then laboratory manager talking already. "I have this very, very important work for us," he said. "It has to be done urgently. I am so busy I do not have any more time to discuss it. Let me know what the arrangements are and when it is finished." He spun on his heels and left, the door slamming behind him. Our mouths opened and closed, and we continued enjoying our break. It was subsequently established that the work was monitoring noise, dust and diesel fumes in tunnels, part of a multi-regional exercise across the network for the Director of Civil Engineering during track repairs and relaying in the open and inside tunnels. In the Midland Region, we had Dove Holes Tunnel, some four miles long, in the Peak District. The work was to be conducted outside normal hours, from 20:00 on Saturday to 08:00 on Sunday and mostly in the winter. Eccentric management maybe but inspirational as well. There was no shortage of volunteers to complete the work, which was done in the allotted timescale.

During a tunnel noise and fume monitoring exercise carried over a weekend, two Scientific Services chemists were leaving Dove Holes Tunnel, their work completed for the shift. It was known that both tracks were in a 'possession' and that no rail movements were allowed until the site was handed back and the equipment train and ballast train had left. With some two miles to walk back to the car, they were about halfway out of the tunnel. It was a very cold January. The tunnel opening in the early morning light could just be seen — a welcome sight. What was not welcome was the increasingly loud roar of an engine behind them. It was a clear instruction that there would be no uncontrolled rail movements during the possession, but something was happening. A light in the distance behind them was getting brighter and closer. When they left the work site, there were locomotives on both tracks, so what was approaching them could be on either side. Fortunately, tunnels are designed to have safety refuges for such circumstances. The railway chemists found one and protected themselves until the locomotive passed at speed, on the opposite track to the one they were on. A close call. It was later established that this particular engine crew had decided to leave because they'd had to turn off their engine to prevent any further diesel exhaust fume contamination and being cold, had decided to leave the tunnel!

The railway industry has long paid close attention to the protection of staff working on the railway infrastructure; safety (other than engine crew) being of paramount importance. Personal track safety training was — and is — mandatory before any access onto the track is permitted and is today more onerous than it has ever been. Railway chemists were no exception, but had they not known what to do and where to go, history could have been very different. Letting a 'bench monkey' out without training could be very hazardous.

Strood Tunnel in Kent, Vince Morris continues, was having the track re-laid, which involved using Drott bulldozers within the confined space and generators used for lighting etc. The possession (the period the track is handed over to the engineers) was from 20:00 on Saturday to 12:00 on Sunday. Because of the potentially hazardous diesel fume concentrations, Scientific Services was asked to send chemists to monitor the atmosphere in the tunnel; it was what we did. The chemists would be on site from 20:00 on Saturday until 08:00 on Sunday, since the final four hours were less reliant on the use of machinery. This was before the time of widespread use of dedicated monitoring equipment using ion specific electrodes and relied on a lot of grab samples of air being bubbled through liquids which changed colour according to the concentration of the nasty gases that were involved. My colleague and I duly arrived on site, set up a base at the mouth of the tunnel and ventured in to take the samples. We had the power to stop work if safe threshold concentrations were exceeded. All went smoothly and at 08:00 we packed our bags and walked off site. On Monday, upon returning to the laboratory, we were greeted by a glowering Area Scientist. "What happened at Strood?" he demanded. "Fine, no problems," was the innocent reply. "That's not what the divisional civil engineer thinks!" The normally calm Area Scientist exploded. "I've had him on the phone since first thing. The job still isn't finished, and the rush hour is in tatters. The trackmen saw you walking off site, assumed there was a problem and walked out themselves. They won't go back until we monitor again and confirm it is safe." "But we agreed that we would leave at 08:00," was the mumbled reply. "Maybe you agreed, but nobody told the men," he said. "In future we — or more particularly YOU — must be present until the job is handed back to the operators." We all learn from our mistakes, which is why I am now a very learned man!

Tunnel inspection trains consist of coaches with roofs modified to give a level walking surface on which inspectors stand (or squat) and tap the brickwork as the train is hauled at walking pace through the tunnel, recording any hollow-sounding areas. The roof was reached by internal steps in each coach. Bo-Peep Tunnel near Hastings has very limited clearance. Two extra rings of brickwork were inserted in the 1860s shortly after it opened, because the contractors building it had skimped on the specification and only built a weak, three-ringed arch rather than the specified five rings. This led to limited clearances and, more relevantly, limited air for the diesel fumes from the locomotive to dissipate. There were concerns that the inspectors may be overcome by the fumes, hence the chemist's presence in a monitoring role. The train had a goods brake van attached as a messing facility. This had the typical coal fired belly stove, and it was roaring away as I joined the train. Monitoring on the roof indicated that fumes were clearing satisfactorily but when we returned to the brake van, staff were complaining of headaches and general nausea. The concentration of carbon monoxide in the van, due to the inability of the fumes from burning coal to easily escape into the tunnel, was several times the recommended safe maximum working level. The obvious remedy until the tunnel mouth was reached was to get everybody up onto the inspection deck where the air was cleaner.

Car fumes

Roll on roll off (RoRo) car ferries, writes Ian Cotter, were operated by British Railways and as the in-house organisation, Scientific Services usually got first refusal to address their problems. The London laboratory covered many of the bigger ports for the cross-channel ferries and thus undertook a wide variety of work for them. A favourite task for staff was monitoring exhaust fumes on car decks, the load space for vehicles, and the fumes produced during loading and offloading, as a build-up of fumes could be dangerous for the crew working there and also for the passengers. Scientific Services staff would be signed on as a supernumerary member of the crew and therefore became eligible for some of the crew privileges. One of these was the purchase of duty-free alcohol and tobacco (long before EU controls came into play). Usually the new supernumerary member of staff went to see the captain to sign on, and most captains would immediately ask if any duty-free was required… which it usually was! Bottles of booze, cigars or cigarettes could be bought at low prices and collected before signing off at the end of the voyage. The day or two away from home was well compensated, and we were looking after the health and safety needs of the staff — which was the main object of the exercise.

This type of work, writes Geoff Hunt, continued with the construction of the Channel Tunnel. The then business responsible for supplying the car carriers wanted to understand what engine exhaust fume levels could be expected when cars and vans were being loaded on and off the new Channel Tunnel rolling stock. These were being built at an engineering works in Lille, France, so a Scientific Services chemist was duly despatched for a one-day exercise (all the client would fund) to monitor exhaust emissions under simulated vehicle loading and unloading conditions. The chemist's remit also included preparing and presenting a report to the French engineering company's board before returning to the UK on the same day.

The first challenge was how to get through HM Customs at East Midlands Airport carrying three boxes of electronic monitoring equipment for carbon monoxide, nitric oxide and nitrogen dioxide as hand luggage, and how to get through the customs at Charles De Gaulle airport in Paris. Letters of explanation in English and French were produced, duly signed off by the Director of BR Research, confirming that the boxes of equipment were scientific instruments and why they were being brought into the country.

We were on a tight timescale. When we arrived at the engineering works in Lille at about 11.00, introductions were made (and translated) and then work stopped for lunch! Our chemist was expecting a short break for a sandwich and then the monitoring exercise could start. Oh no. A full silver service lunch had been laid on by the firm, including copious quantities of wine and French beers! Almost three hours later and dutifully observing his responsibilities under the UK Transport and Works Act, the chemist began the monitoring exercise followed by the report writing and translated presentation to the board and questions. Fortunately, it went well. Then it was quickly into the car for a high-speed drive to Paris to catch the 19:30 plane back to the UK and home.

Confined spaces

Entry into confined spaces requires rigorous examination of the atmosphere within that space to ensure it is safely breathable. Step in once again our brave chemists, with Ian Cotter first on board ship.

One of the many services provided by Scientific Services was to check fuel tanks and wagons for flammable or asphyxiating gases and vapours before anyone went inside to work on them (or on the outside if any hot work was to be carried out). These are what are known as confined spaces, where the gases could not easily escape or were heavier than air, meaning they lay at the bottom of the tank and if any residue remained, could be disturbed just by the act of moving around inside the tank, unsettling the slurry. Many accidents happened across industry when people entered tanks that had not been 'gas freed' before entry, a process requiring ventilating the tank for at least 24 hours and/or forcing compressed air or steam through the space to completely remove contaminants. These days there are strict permit-to-work systems in place for such hazardous environments.

One of the least favourite jobs was when the steam heating coils in the heavy fuel tanks leaked and needed to be repaired on vessels, such as on the MV Vortigern. The Vortigern was a train ferry used to transport freight wagons to the continent and the sleeping cars (Wagon Lit) of the night ferry service. If the Vortigern was out of service for long the night ferry might have to be cancelled, so the work was time critical. The tanks were in what was known as the 'double bottoms' in the bowels of the ship, next to the hull. These would usually be steamed to remove most of the volatile material and then air piped in, but sometimes there was no time to clean the tank entirely and residues would be left inside. The chemist would be kitted out in oil-proof green plastic or rubber clothing and sent in through a manhole cover, pushing the flammable gas detector in front to detect any further problems, having already checked inside before entry — a bit like a canary in a coal mine. This occurred before repair work could commence, to ensure the atmosphere was safe to work in. Once the engineers identified the problem, the defect had to be welded up or plated, with gas measurements taking place constantly. The chemist would then try to clean up and take the trip home by train, often smelling of crude oil.

Similar circumstances applied at the Engineering Development Unit (EDU) at the BR Research site in Derby, recalls Geoff Hunt. Tank wagons would come into the workshop for repair or installation of sensors and so confined space testing was required before any work could be undertaken. When the purging was expected to be completed, a chemist was summoned to carry out tests for flammable gas content. Often the wagons that came in had the entry port on the top, which necessitated the chemist having to climb the boarding ladder to the top of the tank, with his equipment, and then walk along the walkway to the entrance port. You needed a head for heights and be as nimble as a mountain goat!

Treating an empty wagon as a confined space seems a bit extreme, writes Vince Morris, but if the cap fits... Darnall wagon repair workshop, on the outskirts of Sheffield, unsurprisingly carried out wagon repairs. One of the routine overhauls undertaken was to coal wagons, and one of the requirements of the overhaul was to clear the drainage holes in the wagon floors, which often became clogged with coal dust, corrosion products and general debris. If necessary, the process was carried out using an oxy-propane flame torch to burn out the muck, usually with the fitter standing inside the wagon. The flame torch consisted of two streams of gas, oxygen and propane, which are combined at the torch head and when ignited burn with an intense flame. The oxygen and propane are supplied to the head via tubes from gas cylinders, typically kept some distance from the flame. There were control valves on each cylinder and on the torch head itself.

On this occasion, the fitter took his propane torch into the wagon but was called away. Unbeknown to him, although the oxygen and propane valves on the torch were off, the valves on the gas cylinders on the ground outside the wagon were still open. There was a very slight leak at the torch valve of the propane line and while the fitter was away, the wagon slowly filled with propane gas from the leaking valve. Since propane is heavier than air, the gas would normally have escaped from the wagon via the drainage holes. But the very reason the wagon was being repaired was because the drainage holes were blocked. When he returned to the job, the fitter climbed back into the wagon, turned on the propane gas at the torch and struck a light.

The resulting explosion blew him out of the wagon, and he was severely injured. Explosions only happen if the ratio of flammable gas to air is within certain limits, i.e. below the Lower Explosive Limit (LEL) when too low a concentration exists and the propane would not explode due to a lack of the gas; and above the Higher Explosive Limit (HEL), when there is too high a concentration and the propane would not explode due to a lack of oxygen. The fact that it exploded was an unfortunate circumstance of timing, in that the air/propane mixture was within the explosive range (between two and ten per cent propane). Thereafter at Darnall, the open wagons were treated as if they were a confined space and tested before any entry and works were permitted.

Stinking brakes

What is that stink when a train brakes? It was a common question asked in the 1960s to 1990s by passengers and railway staff alike, writes Dave Smith. Letters arrived in Scientific Services by various routes asking what was that strong acrid smell in stations when trains braked approaching the platform, and also asking if asbestos was present. It was therefore essential to be careful in responding to such enquiries in case it was from the media looking for a story. So not only did we have to be technically competent, but we also had to be diplomatic and cautious in any response (not something that came easily to some!).

As chemists we were loath to publicly announce that one of the ingredients of the smell was in fact hydrogen cyanide, as this would no doubt lead to dramatic headlines in certain tabloid newspapers, renowned for their extreme scientific credentials. An enquiry from an environmental health officer in Leeds threatened to have all his thirty officers board every train passing through the city to test what the smell was. Fortunately, we too had developed equipment we could place on trains and collect samples of the braking emissions, which were then analysed in the laboratory.

Everything has to be in context and although hydrogen cyanide was a by-product from the train braking process (and it is possible it was an emission from car brakes too), the concentrations were very low — less exposure than you would get when you open a pack of almond nuts. This led to the question: should we be banning packets of almond nuts and railway braking systems as they stood at that time?

Asbestos exposure from brake block emissions was always a concern with railway employees and the travelling public. This was not helped by the odours released during braking and although it is well known that asbestos has no smell, it was easy to conclude that the acrid aroma could be an indication of harmful contaminants. Consequently, numerous tests were undertaken on a wide range of rolling stock to identify if there were any airborne asbestos fibres released during braking activity. The results proved this was not the case. What really happened when brakes containing asbestos fibres as a reinforcement to hold the pads together were applied was that the high temperatures generated converted the white asbestos (chrysotile) to a crystalline material identified as Forsterite, which was not a hazardous fibrous form in the same sense as asbestos. The smell was in fact the combustion of the resin and binder from the brake pad as it heated up.

In one reply to a letter of complaint regarding smells from braking from a doctor, we told him that there was nothing to worry about. We had analysed the fumes and considered there was no health hazard. He responded with a very frosty letter, advising us that he was the Chief Medical Advisor to the Government and did not expect to receive such a patronising response.

Passing Signals at Danger (aka SPADs)

Dave Smith outlines two instances when thinking you are doing the right thing can make matters worse.

One of the most interesting aspects of our analytical work concerned investigating incidents and accidents to help determine their root cause. Trains passing signals or crossing points when they were supposed to have stopped were obviously big issues.

Track circuits and signalling systems relied heavily upon the use of relay switches. These are small electrical switches with silver contacts contained in a sealed plastic box. They were supposed to fail safely, setting the signals to danger even if there was no problem on the line itself. This was achieved by the silver contact points coming together if there was a fault in the system. The relays were enclosed in boxes to prevent rain or other contaminants entering them and shorting the contacts. In the early days we found that the very act of sealing them could cause problems. The plasticisers in the plastic casing migrated and sometimes

133

coated the contact points with an insulating film, so the contacts could not fail safe. Many studies were carried out to determine which materials should be used and to minimise such risks, but a number of accidents occurred because of what was termed wrong-side failures. Much work was done to determine the cause of the problems. Electron microscopy, scanning electron microscopy and infrared analysis were commonly used to identify the small amount of contaminant on the contact points and whether it was organic or inorganic in nature.

The lack of air flow within the boxes was considered to be one cause, allowing the build-up of plastic fumes which subsequently condensed on the relay terminals. Some boxes had small vents with a very fine mesh over them, allowing any vapours to escape. Unfortunately, these also led to a new problem: the very fine mesh did not prevent the ingress of very small insects who seemed fatally attracted to the contact points. Another problem-laden solution to protect them from degradation and insect's ingress was for engineers to encase the relays in silicone after installation. The trouble was that the silicones used at that time were cured by a process which emitted acetic acid (which makes up five to eight per cent of vinegar) vapour. This in turn attacked the wiring and chips. The engineers were advised to put their vinegar on potato chips in future!

We decided to develop a method for determining what materials could be used which would not degrade and emit vapours that attacked wiring and silver contact. The method entailed subjecting the individual materials in a small sample bottle to an extreme range of temperatures and extracting any gases emitted. These gases were analysed by mass spectrometry and those that gave off emissions thought likely to attack components were rejected. A database was built up of acceptable polymers, paints and so on. This work brought us to the attention of the European Space Agency, as they saw the need for better evaluation of materials used in similar critical locations.

Safety device failure on locomotives

We were confronted by an engineer from the Director of Mechanical and Electrical Engineering (DM&EE) department with an orange box under his arm. The box had a maximum dimension of about 30cm (one foot in old money). It was described as a Driver's Safety Device (DSD) and we were told it was filled with electronic and electrical devices which applied the brakes of locomotives should the driver fall ill and be unable to drive the train.

Apparently, it had failed. The box was opened and inside there was indeed lots of electronic gear and wires, but which were covered in white crystals and corrosion. It was obvious that the components had been attacked by something contained within the box. The engineer was not too happy with the response that we could probably get six PhD papers on trying to understand the chemistry of what had gone wrong. He wanted the main problem identified quickly.

To resolve this, we used a wide range of equipment and analytical techniques to identify the various crystal growths. After a few days we concluded that the dominant problem was that gases emitted from the orange alkyd paint had attacked the components and given rise to the white crystals, resulting in the failure problem. As was expected, the engineer did not want a detailed analysis of what was in the box — he wanted his box working. All the evidence pointed to the conclusion that the paint used inside the box was still drying by the time the box was sealed, and solvent fumes were attacking the insulation. He was advised to use a different surface coating, one that was fully cured before the internal components were fitted and the box was sealed. There was a cautionary note added to the effect that this was just the solution to the immediate problem and other processes may cause further problems if a more comprehensive analysis was carried out.

A year or two later, an engineer from DM&EE turned up with an orange box under his arm. The boxes were failing again. This time the source of the problem was different as they had changed the paint but were again using unsuitable materials within the sealed unit. Our report made it clear that they needed to recognise the selection of inappropriate materials was a design problem. We offered to give a free presentation to the designers about the perils of placing electronic or electrical devices in sealed boxes, based partly on our work for signal engineers on wrong side failures in signalling relays, and our recognition by the European

Space Agency as experts on how to test paints and plastics.

A few years later, a private engineering company submitted a sealed device exhibiting similar defects but was not a DSD but a control panel on the new coaches operating on the East Coast mainline. The same problem of not recognising that paints and plastics emit damaging fumes was diagnosed, and the same statement that it was a design problem because the engineers failed to appreciate the chemistry of the materials they were specifying. We were subsequently asked to attend a meeting with senior DM&EE management. Unbeknown to us, the equipment we had received from the engineering company was for use on trains and the DM&EE had supplied the specification.

Why, they asked, had we not informed them of the problems in the selection of materials for sealed units? At no time had we been approached by the DM&EE and as far as we were concerned, we were dealing in confidence with a private customer (which in the run-up to privatisation we were told to do). They were still not satisfied and so I produced two reports and and pointed out that we had offered to provide a training session on the nature of the problem in design. There was a moment of quiet, and then we were asked to leave the room and nothing more was heard about it. We subsequently learnt that the independent client's component was from the then new electric stock used on the Kings Cross to Edinburgh line, and its failure could allow doors to open at 100mph! British Rail was holding the supplier of the rolling stock accountable, but it seems they specified the safety device used and the cost of modification was stated to be millions.

Health and safety management is high risk

The lessons to be learned from the following sad reflection makes you wonder why we bother trying to be safe. Not many of the approximately 800 staff in BR Research were chemists, but many of them used chemicals.

In the 1970s, Scientific Services was confidentially advised that a member of staff had unsuccessfully tried to end their life using some of the chemicals too readily available in their department. At about the same time, a researcher, who had some very strange ideas and was annoyed at the lack of career progress, made serious threats to colleagues and management. A bottle of arsenic was subsequently found in her desk drawer which did not relate to any activities taking place.

The trouble was that when a research project finished, the partially used bottles, cans and drums containing chemicals were virtually abandoned. Over time they could become unstable and the label could drop off or become unreadable — and then they obviously posed a risk.

The Derby laboratory was then a small department and part of the BR Research site, so funds were sought to enable us to collect all unknown materials and dispose of them appropriately. This idea was readily accepted by senior management and from then on, we had deliveries from everywhere on the site. Alan Brown, a very able Scientific Services chemist, was assigned to the identification and disposal of these mixed bags of chemicals. Much that arrived was easily sorted and could be disposed of in a safe manner.

One day there was pandemonium as Alan ran down the laboratory shrouded in a white cloud. Several fellow chemists went to his aid and ventilated the laboratory. The result was that four of us ended up in the Accident and Emergency department at Derbyshire Royal Infirmary for an overnight stay for observation. Alan, who had been more significantly exposed, was despatched to Derby City Hospital for several days. Apparently, all Alan did was visually examine a bottle shortly after it arrived and gently tapped the stopper. It exploded and the titanium tetrachloride inside — which is highly unstable — resulted in him being showered with hydrogen chloride, or hydrochloric acid. Following an intensive investigation, breathing apparatus was sited at both ends of the operational laboratory just outside the main doors. All laboratory personnel were trained in its use and could rescue other individuals if necessary. Inside the laboratory, simple ori-nasal respirators were installed on the central pillars as a first-line emergency response if a similar accident occurred.

One day, a safety audit was undertaken by the research safety officer. He asked: "What is this device on the wall?" He was told it was breathing apparatus and why we had two such kits. He then asked: "How do you know they will work?" We provided certificates that showed the kits were routinely serviced. "Who is trained?" was the next question, and after the response "All", he decided to ask the most convenient member of staff about the kit and also to demonstrate it.

All had gone well so far... but then he chose Glynne! Now Glynne was indeed trained and a very competent chemist, but he seemed to have a far higher number of laboratory disasters than the other chemists and it was anticipated that the kit would be used for rescuing him rather than him rescuing anyone else! The auditor went through the protocols of confirming the earlier responses and asked Glynne to demonstrate the apparatus. He opened the box and put his arm into one of the cylinder straps, over-enthusiastically swinging the cylinder round to put on the other strap. The cylinder went higher in the air than intended and hit Glynne's skull with such force that we had to call an ambulance. Yet another accident enquiry.

We all survived long enough to write about our experiences.

CHAPTER 7

Fire Technology

It took a fatal fire to shake up BR's response to fire safety, but we found that, rather than lagging behind industry in many areas, we were in the vanguard. Here we examine the Scientific Services approach to fire testing and the investigation into several notable fires. The subject is introduced by Vince Morris

A TRAGIC EVENT led to Scientific Services having to learn some new skills. Eleven people died when a fire broke out on the Penzance to Paddington sleeper service as it approached Taunton on 6th July 1978. The early investigation by Dr Duckworth, of the Home Office Forensic Service, was supported by Alwyn Platts, of Swindon, and Jim Ward, who eventually became Head of Scientific Services. They quickly discounted the possibility of arson and showed that the ignition of used bedding stacked adjacent to a heater was the reason for the tragedy. The heater had recently been converted from steam to electric as part of an improvement programme. The deaths were mainly due to the inhalation of smoke and toxic gases (see photo 53).

Questions were raised in the Houses of Parliament about the fire safety of rolling stock and in response, British Rail Research decided to set up a Fire Technology Team — and where better to put it than within Scientific Services. Dave Smith was recalled from leave and asked if he would take on the responsibility of acting head, which he fulfilled until Vince Morris took on the permanent role early in 1981. The main role of the team was to respond to the general malaise within the fire research community, as exposed during the Taunton inquiry, by relying on tests involving the exposure of a small sample of material, in isolation, to a flame, and attempting to extrapolate the results to the behaviour of full-size components in the complex surroundings of everyday situations. From the start we looked at real items, such as seats, in the orientation in which they would be used in service and, where possible, with knowledge of the fire behaviour of surrounding materials, such as paint, adhesives, flooring and panelling. We were looking not just at ease of ignition but at smoke and toxic gas production, and the effects of air conditioning and so on. The safety role of the team, as we saw it, was to keep the goalposts the correct distance apart. Too close and they would touch, letting nothing through on safety grounds. Giving passengers concrete seats may prevent a fire but it would also discourage them from choosing to travel by rail. If the goalposts were moved to the edge of the pitch, we would score own goals by abandoning the concept of safety and taking a lax attitude to our passengers' well-being.

Testing times

This was the time when new sleeping cars were being built and standards on bedding and fixtures needed to be determined. Modifications had already been proposed to the method of construction, using metal as a fire barrier between the berths. Although Scientific Services had some experience at setting fire to things and the odd explosion, albeit usually by accident, the skill in fire technology other than arson was zero. There needed to be a fast learning curve.

At about the same time there was a tragic fire at Woolworths in Manchester and the Fire Research Station based at Borehamwood, near Elstree Studios, was asked to help in the investigation. We were privileged to be involved in a minor way, in watching and helping in the investigation and learning the skills of fire technology at the same time. Should one ever pass through Bedford by train, there are two imposing buildings in the distance to the east. They are in the village of Cardington and are the two hangars built for airships. One hanger was used by the Fire Research Station for carrying out full-scale fire tests. It is an enormous building and a detached house, built for a fire experiment, was dwarfed by its surroundings. When the klaxon went to announce an explosive or fire test, bicycles were available to enable a reasonably quick exit before the doors closed. The simulation of the fire at the top floor of Woolworths involving the furniture was amazing and frightening. Once the foam soft furnishings caught light, the fire spread rapidly, with copious black smoke egressing the floor space set up to simulate the original layout. It nearly caught the camera crew by surprise, and they had to be told to move as they were clearly spellbound by the fire. The result of this experiment very much influenced the Government in introducing the standards for our chairs and sofas in use today, with the familiar cigarette and match labelling.

British Rail Research had a Plastics Development Unit (PDU) and there was also a Textiles Unit within the Mechanical and Electrical Engineering Department, responsible for the specification of seating materials. Alan Blount from PDU was involved in evaluating the foam used in vehicles and Harvey Jones from the Textiles Unit supported Peter Davies and Dave Smith in the Fire Technology Team in establishing full-scale tests of bedding and seats. This was to ensure that future furnishings on rolling stock did not behave in the manner experienced on the train at Taunton or like the sofas at Woolworths (see photo 52).

The need for realistic surroundings for testing bedding and seats on trains was recognised, since the confined space and low ceilings on a train posed far more risk of harm from a fire than in a building. A fire test facility was built within a compartment of the railway coach with observation space and the ability to escape if a test went wrong. The coach was immediately recognisable because it was the only one on British Rail fitted with a chimney. Like several research department vehicles, it had a name painted on its side. We had, for obvious reasons, chosen the name Phoenix; luckily, we were able to stop the sign writers spelling it Fenix, which was what was written on the pad after our tele-phoned request for them to undertake the job. The modifications were basically to completely line the compartment next to the end vestibule of the coach (the so-called fire cell) with non-flammable cladding and a door leading out to a corridor to allow smoke and toxic gases to flow past monitoring equipment.

To ensure we knew the height of the smoke and gases as they traversed the corridor (relevant since it was dependent on the temperature of the burning sample under test) and we used a baffle, somewhat like a perforated egg carton, which allowed us to measure various densities and concentrations at different levels. The smoke finally exited through the open doors of the guard's compartment at the far end of the coach, which was also from where the air to support the fire was drawn into the fire cell, so the egg carton also acted to separate incoming and outgoing gases. We were positioned in the vestibule behind a non-flammable barrier (with door access to the fire cell), with a polycarbonate panel that allowed us to watch and film the progress of the fire. We had fire extinguishing equipment with us when testing seats and mattresses and wore fire resistant Proban-treated overalls and breathing apparatus. The fumes from the polyurethane contained high levels of cyanide and carbon monoxide, and Proban-treated clothing was essential when extinguishing the fires. We had many a scare during testing: when a mattress was tested and dowsed with water it was considered safe to enter the test cell. Even though it was well-watered, the foam suddenly reignited and the mattress was soon ablaze, causing us to run to safety. And when a misdirected jet of extinguishing water landed on a floodlight installed to allow satisfactory filming, the resultant noise of the exploding bulb had the entire team flat out on the floor in a micro-second. Phoenix was initially stabled at Mickleover, Derby (the location of the BR Research test track) but was subsequently moved (with the chimney lowered!) to a siding at the bottom of the Research Division site's on London Road, Derby, near where all the new stock built by BREL in the adjacent Carriage and Wagon Works was taken out onto the main line.

Eventually a custom-made brick building was built to replace Phoenix. It was designed by a BR architect with the surname Dante, so of course the building was dubbed Dante's Inferno! It was in the yard beyond the Derby laboratory buildings, and we approached Derby City Council to get their reaction to the possibility of pollution resulting from our activities. The reply took the view that the railway was already so polluting that a little more would not make much difference. The results of the work led to the development of a specification ensuring BR had robust requirements for fire safety for seats and mattresses. We also tested the wooden and plastic panelling, and associated decorative finishes to be used within vehicles, again at full-scale and in the correct orientation. One success of such testing, on both panelling and seating, was that we were able to demonstrate to suppliers the shortcomings of their products in a 'real' fire, who had previously relied on the results of small-scale tests. This led to several reformulations, and Scientific Services was approached for approval of materials prior to them being offered to our and other industries. One unexpected enquirer was the Prison Service, which wanted to upgrade cells at Broadmoor and decided they were very similar to railway compartments!

Another development arising from our testing was that the standard fire test on a seat — which involves using a seat back and squab (the bit you sit on) positioned in an open frame at right angles to each other, and using a wooden crib to simulate four sheets of newspaper as the fire source — did not represent a real world situation. If, instead of the frame being open as in the test for domestic furniture, we put sides on it as it would appear in a railway coach, the burning characteristics were entirely different since the air flow was modified. We therefore recommended that arm rests on seats should be solid rather than open: a simple design requirement which considerably increased fire safety. This was, of course, only applicable in the railway (and possibly airline) situation, where we could specify the positioning of the seats as well as their materials of manufacture.

Vince's takeover period from Dave Smith was enhanced by events in Camden and by the building/fitting out schedule of the Mk3 sleeper to replace the Mk1 stock involved in the Taunton fire (there never was a Mk2 sleeper). Due to the need to not delay the introduction of the new sleeping car fleet, the gut response to the Taunton fire — undertaken prior to our involvement — was to re-engineer the Mk3 build by inserting additional metal flooring and partitions, thus considerably increasing the weight of the vehicle. Although tests we subsequently carried out showed the vehicle was the most fire-resistant piece of rolling stock ever produced, it came at the expense of operating efficiency with tons of extra metal being dragged around the country every night. Some of the understandable modifications were arguably 'over the top'.

Of course, testing in any form is unlikely to replicate a real fire, so fire investigation was still very much on the agenda for the Fire Technology Team, dealing with both accidental and intentional fires. Here, Dave Smith recalls a fire on a train at West Kirby, near Liverpool, on 28th June 1980. A class 503/1 power car M28683 was gutted within two to three minutes. Had the train not been stationery at the station, the fire could have resulted in casualties. A serious electrical fault in the distribution panel fuse box was the source, while the rapid spread of flame that engulfed the vehicle was attributed to the painted plywood ceiling panels.

The fire was so serious that the stock was considered high risk, and immediate action taken to prevent a recurrence of the electrical fault. However, the spread of flames was also considered high risk and we were asked to determine if it was possible to improve the fire resistance of these vehicles. In order to do this, small-scale tests were carried out but a full-scale test was essential to ascertain the effectiveness of any remedial alteration to the vehicle.

The initial lab work, concentrating at looking at the ceiling panels, found they had about 20 layers of paint and varnish on a non-fire retardant plywood. The vehicles were of pre-1940 vintage and the layers of paint corresponded to the surfaces being re-coated every few years. A simulation using a 1m2 panel from the vehicle and a gradual build-up of the fire to a similar energy to that in the vehicle showed that at just 60 per cent of the calculated maximum intensity experienced on the train, there would be a rapid escalation of the fire, probably similar to that observed at West Kirby, after four minutes. In a subsequent test, on an identical panel with an intumescent paint, which swells when it gets hot and thus produces an insulating layer between the plywood and the flames, the fire was arrested before it could spread.

However, it was necessary to demonstrate that the application of intumescent paint as a treatment to prevent the rapid spread of flame would behave in a similar manner when present in an actual vehicle. To do this, we needed actual rolling stock and some clever technology to record the fire as it developed using instruments, gauges, cameras and so on.

We were provided with two similar vehicles (502/503) and set up a full-scale test in Chaddesden sidings at Derby. This was an ambitious large-scale exercise. A marquee with TV monitors was set up so dignitaries from the Ministry of Transport, Fire Research Station, Railway Inspectorate, the fire brigade from Liverpool and railway engineers could witness it (see photo 48). We had made use of instrumentation specialists and engineers to provide a superb test bed. It was our understanding that if our treatment did not work, the vehicles would have to be withdrawn.

There were just two of us at the business end, with the fire brigade on standby to quench the flames when the expected flash-over occurred. Peter Davies gave the commentary and I operated the heavy video recorder. In those days, equipment was expensive, and we could not justify using cameras that might be destroyed in the fire.

The first test was unbelievable! A wooden crib made to a BSI specification to produce the equivalent energy to that when the fuse box failed was installed in the corner of the vehicle against an end bulkhead (see photo 49). This was to ensure we had the worse-case scenario. The crib did not produce the rapid energy flash that was experienced at West Kirby but built up slowly and enabled us to observe the various stages in the fire's development. The fire proceeded as expected in the untreated vehicle and after about four minutes, the ceiling panels caught fire. A rolling ball of flame swept across the ceiling towards the two-man film crew sited near the doorways (see photos 48 and 50). The filming and commentary were orderly until then but as the fire increased Peter Davies understandably forgot himself and quite sensibly shouted: "Let's get out of here" — with a choice swear word inserted. This bit of tape had to be censored, so the roar of the fire was lost for a few seconds. That part of our risk controls worked well! More tests were carried out with just the ceiling panels treated with intumescent paint, and one with both the side and ceiling panels treated. The outcome was most positive, and the West Kirby units continued to operate, albeit with unfashionable ceiling paint (see photo 51 - prior to test).

Here Dave recalls a serious fire on a tamping machine near Pilmoor, Yorkshire, at 03:55 on 2nd April 1981 — which could have easily killed three railway workers. A tamping machine is an expensive, complicated piece of equipment used for maintaining railway track. In simple terms it lifts the rails and sleepers and tamps the ballast into place. This means the track is restored to its correct level after being disturbed by track movement, which occurs over time due to the substructure and traffic speed and frequency. Tampers normally operate at night when there is little traffic. The machine in question (Number 73432) used a 230bhp Rolls-Royce engine for moving the vehicle and powering a hydraulic system used to lift the track and tamp the ballast into place with tines.

The diesel engine was located in a housing directly behind the cab, where the driver and other operatives were occupied running the vehicle. A massive fire vented out of the engine compartment directly onto the cab, quickly causing severe damage. The configuration was such that it acted like a giant flame thrower. Had the crew followed procedures, which necessitated that they had a chain across the open doorways, they may not have escaped unharmed. As far as could be determined, the lack of following safety procedures saved their lives (see photo 46)!

Scientific Services at Doncaster were involved and Dave (as acting head of the Fire Technology Unit) met Bob Symcox from the Doncaster laboratory on site with representatives of the machine manufacturer, railway officials and the Principal Railway Employment Inspector of the Railway Inspectorate (part of the Department of Transport). The investigation was entirely focussed on the engine compartment. The general view was that there had been a fracture of a diesel pipe creating a mist that was drawn through the compartment by the cooling fans and impacted directly onto the hot exhaust pipe outside the engine compartment (some of which had manifolds of such poor quality that they would draw gases into the hot exhaust stream within the pipe). However, the Scientific Services approach was more independent, and we found a fractured metal hydraulic pipe which could have led to a mist of hydraulic oil, rather than diesel, being drawn into the engine compartment (see photo 47). This was backed up at the subsequent inquiry, when the look-out man said he had warned the crew of the hydraulic leak prior to the fire. Although only a possible cause, we made enquiries as to whether there had been similar incidents around the country in any of the other 30 similar vehicles.

It was found that a similar situation, but without the same consequences, had occurred on a tamper in the South West and there had also been an event on a rail regulator (another piece of equipment for track maintenance using hydraulic systems) arising from hydraulic spray impinging onto the exhaust system. In the South West they replaced fixed metal hoses by more flexible hoses that would not fracture in the same manner. Our investigations were not welcome as they demonstrated that the problem was known but that little or no attempt had been made to tackle it throughout the fleet of vehicles. I even received an anonymous phone call advising not to continue this line of investigation. Braving mafia-style attacks, I submitted a report to the inquiry outlining the possible scenarios and suggesting mitigation techniques. Extracts were included in the official Department of Transport report.

The investigation of a fire at Camden unofficially marked the handover of the reins of the Fire Technology Team from Dave Smith to Vince Morris, and now he takes over the story.

Camden, in North London, has railways links going back to the earliest days. A goods yard was established there in 1839 by the London and Birmingham Railway, that location chosen because of the proximity of Regent's Canal. Over the years various railway buildings were constructed on the 25-acre site and even today, despite selective demolition, it probably has one of the highest concentrations of former railway warehouses in the country. One building is missing, however. In the early 1980s, the General Manager's (GM) Warehouse was destroyed in a fire reputed to be the biggest blaze dealt with by the London Fire Brigade that year. Dramatic video footage obtained from a nearby nightclub captured the entire brick facade of the building falling to the ground. Fortunately, there were no casualties.

Being a British Rail-owned building, it was under the jurisdiction of the BTP. Officers from Euston were tasked to investigate the fire, more, I suggest, in hope than expectation of discovering the cause, let alone to establish who, if anyone, was involved in the ignition. They asked for assistance from Scientific Services staff.

It was a massive fire. The warehouse had largely outlived its original purpose and although the GM still stored an assortment of furniture in the building, large areas of it had been into small offices and storage areas rented out by the British Rail Property Board (BRPB). There were dozens of tenants, most of whom would, as a result of the fire, be claiming against BR. The multi-storied building was now a burnt-out shell, and debris from some of the upper floors had fallen to ground level. However, in so doing it had protected items in the lower floors from being fully burnt. It was possible, by hard slog on the BTP's part, to cross reference with BRPB as to who rented what and obtain from tenants a list of what they had. By assessing the damage, we established on which floor the fire broke out. It sounds simple but it was a long, drawn-out process, both for BTP who experienced resistance from tenants as to precisely what they were doing, and for us having to sift through tons of debris identifying burnt remains and placing them in 'floor' order. We were also trying to work out the area of most fire damage, which would indicate where in the building — which had a floor area of several hundred square metres — the fire had started. It was a bit like three-dimensional noughts and crosses: we were trying to pinpoint not only on which floor, but also whereabouts on that floor.

After numerous visits and much speculation, we had a breakthrough. We located a carpet roll which, although on its side when discovered, had a burn pattern suggesting it had been stored upright at the time of the fire; the top of the roll was severely damaged but the bottom (or what we assumed to be the bottom) had survived relatively intact. This suggested the carpet had been on the storey where the fire started, fell through the floor when it gave way, landed on its side and stopped burning. But why had the floor failed if the base of the carpet had survived? Further examination of the debris in the area of the discovery suggested there had been several carpets in the vicinity. We speculated that this particular roll had been surrounded by others which protected it from burning at the bottom because of the lack of free air to support combustion, and that the others had burnt sufficiently to cause the floor to fail. The fact there was burning within the roll suggested the fire had started or been started not at floor level, but at some height above — which was itself strange. There could be several reasons why the fire within the recovered carpet had not reached the bottom of the roll: lack of air or some mechanical restriction. One possible explanation was that a flammable liquid had been thrown over the carpet rolls and ignited but if we were right that this roll was surrounded by others, it was possible that liquid had only splashed down the roll and the burning represented where the liquid settled.

Our findings were only speculation and would not have stood up in court, however they gave a pointer to the investigating police officers as to where to start looking and who to investigate further. The tenant in the location we had pinpointed was identified and interviewed. From memory (and of course I was not at the interviews), he described himself as a general dealer and admitted he may well have had carpets in his storeroom. Then the Property Board came up with an interesting fact. Every year when the rent was due, this tenant counter-claimed against BR for damage to his stored items, and the claim neatly matched the rent due — resulting it in sometimes being waived. By coincidence, one of his claims was for water damage to musical instruments in his possession, and the Property Board had asked Scientific Services staff at Muswell Hill laboratory to investigate. I fished out our report. Although the examination confirmed water damage, there was a paragraph to the effect that it had not been possible to determine where the water had come from. The revelation of almost annual damage claims set alarm bells ringing, and the BTP established that in addition to claiming from BR, which carried its own insurance, the tenant also claimed against his own insurer. So, although the tenant could not be prosecuted for arson (and it would be impossible to investigate any prior claims against BR), he was successfully prosecuted for fraud. The fire investigation allowing a suspect to be identified was not in vain.

An amusing side issue was that an anonymous letter was received by the police suggesting a local homeless man had caused the fire. The letter was submitted to the London laboratory for examination using the recently acquired ESDA machine (electrostatic detection apparatus), which can pick up indented writing. The letter had been written on notepaper torn from a notepad and the address of sender, which had been written on the previous sheet, was clearly revealed. An officer went to the address and the surprised occupier admitted she had written the note with the intention of getting the homeless man, whom she disliked and was afraid of, arrested. She asked the officer how the police knew she had written it. If we believe him, his reply was that in important cases, officers sit inside every pillar box in the area and intercept letters addressed to the police and follow the person who has posted the letter. Apparently, she believed him! I'm not sure I did.

Another investigation I recall involved a fire on 16th August 1983 on the 18:30 Edinburgh push-pull service to Glasgow as it approached

Cadder. Due more to luck than judgement, no one was seriously injured but the coaches concerned were to the latest Mk3 design as used on the HST sets, so there was considerable concern within the railway industry and the Department of Transport about the causes and spread of the fire. It gave a boost to the development of a British Standard for fire safety in railway vehicles, setting requirements for the selection of materials and design which became mandatory for all new vehicles supplied for use on BR. Back to Cadder, on a warm summer's evening. It had been very warm and dry in Scotland for the previous month and this was a significant factor in the fire. As the train approached Lenzie station at about 90mph, the communication cord was pulled, and the driver cut off traction power. The train came to a stand just short of Cadder signal box some five miles from Glasgow. Looking back along the train, the driver saw thick smoke coming from the fourth and fifth coaches (of a five-coach train, but with the propelling locomotive behind them). The passengers managed to get out of the train and onto the track and, luckily, since it was so close to the signal box the signaller saw the problem and halted all traffic. The passengers were questioned by BTP officers both at the scene and subsequently, and a picture emerged of a developed fire suddenly bursting into the vestibule between the two coaches: spreading into one through the vestibule door which had failed in the open position during the journey, and via the ceiling into the other. That the fire was described as bursting into the vestibule suggested it had been burning outside the train, which seemed a bit unusual but also gave a clue as to how it could have started. The affected coaches were taken to the Glasgow workshops and examined by Roger Hughes from the Glasgow laboratory. He concluded the fire had started in the foam gangway connection between the two coaches. The foam was meant to be protected by an integral PVC skin, but examination of other coaches showed this skin was liable to splitting. There was a small gap between the bottom of the foam and the tread-plate of the corridor connection, probably larger than designed due to wear from motion of the train.

Roger suggested a discarded cigarette had been dropped in the vestibule, rolled into the gap and then worked its way to the outside of the foam, where the air flow induced by the train movement caused it to flame and ignite the foam. He demonstrated this process in a small-scale laboratory experiment. This mechanism was accepted by the Strathclyde Police forensic team, who was called in to investigate the incident as a potential arson. Roger also contacted me as the Scientific Services fire 'expert' (a horrible expression!). Two questions remained: why was foam used for the gangway connection when they were previously rubber bellows, and why had there not been previous serious fires if the foam ignited so easily? (One had occurred on an HST in 1981 but was extinguished by the guard).

Both had simple answers but show how good intentions can have serious consequences. At 23 metres, the Mk3 coaches were the longest ever to be proposed for use on BR (the longer 'twin sets' used pre-war on LNER were articulated above the centre bogie), and the designers were concerned that, on tight bends, the rubber bellows might pinch passengers so opted for the customer-friendly soft foam. However, as indicated, the foam did not wear well and was being replaced by bellows as they went through works. But why no previous significant fires? It had been a dry summer and the Scottish Region (BR formed ScotRail the following month, September 1983) decided that, to save water, they would not pass trains through the washing plants. It was established, once the drought was over, that when washed, the foam absorbed large volumes of water and typically became saturated and would not burn.

The inquiry into the fire was chaired by Lieutenant-Colonel Tony Townsend-Rose, of Her Majesty's Railway Inspectorate (HMRI), who recommended, during the hearing, that all foam gangways should be replaced urgently. As an indication of the notice taken of HMRI recommendations, ScotRail stopped and replaced all remaining foam gangways — 19 in total — before the report was issued. He also recommended that escape plans should be displayed in all coaches, and the now-familiar safety notices date from this time.

There is an amusing personal aspect to this fire. As I noted previously, the Cadder fire gave a boost to the production of a British Standard on fire safety on trains. The process began as a result of the Taunton fire, and Lt. Col. Townsend-Rose was the Railway Inspectorate representative on the committee drawing it up. I was the BR representative, and there were committee members from London Transport, the Government's Fire Research Station and the manufacturers. As a result of previous meetings, I knew Lt. Col Townsend-Rose quite well. I was invited by Roger Hughes to see the damaged coach in Glasgow Works. I'd had a meeting in London and booked a berth on the sleeper from Euston. By coincidence, Townsend-Rose was going to Glasgow on the same train, at the request of the Scottish Region's management, to have a pre-inquiry inspection of the damaged rolling stock. He had noticed my name on the sleeper berth register and met me on the train.

When we arrived at Glasgow Central the next morning, we walked along the platform and were approached by a uniformed chauffeur who asked if my companion was the Railway Inspector from London. Townsend-Rose said he was, and the chauffeur said breakfast had been arranged at the Central Hotel. Afterwards he would drive the inspector to the works to inspect the coaches and then back to Scottish Region HQ for lunch with the board. Without batting an eyelid, Townsend-Rose said: "I presume the invitation extends to my technical adviser who has come up with me." What could the chauffeur say? I had breakfast, inspected the coaches with an awed works manager and a silver service lunch with the entire hierarchy of the Scottish Region, who were doing their best to impress the inspector. I was very much an afterthought: a printed menu produced for the senior officers' mess was headed: "Programme for visit of Lt Col Townsend-Rose, HMRI, to the Scottish Region Board. Thursday 25th August 1983." Below his name had been hastily typed: "And Mr Vincent Morris, Railway Technical Centre."

Probably one of the most spectacular fires in peacetime Britain occurred on 20th December 1984, when 600 tons of petrol in ten tanker wagons caught fire in the depths of the trans-Pennine Summit Tunnel, near Todmorden, in West Yorkshire. The brick-lined tunnel, built by George Stephenson and opened in 1841, is 2,885 yards long and on the route of the tanker train from Haverton Hall to Glazebrook. The train entered the northern end of the tunnel just before six in the morning and operated the treadle mechanism, which warned the signalman at the other end that the train would exit and pass his box in a few minutes. But the train was in trouble. A bearing on the fourth tank wagon was running hot and eventually collapsed, causing that wagon to derail and take some of the trailing wagons with it. There were several fires caused by leaking petrol coming into contact with both sections of the failed bearing: one still on the wagon, the other by the track side some distance back, since the wagon had been pulled forward before it came to rest.

At this stage, coincidences get out of hand. One fire was in West Yorkshire and one was in Greater Manchester; the signalman who had registered the train entering the tunnel went off duty at 05:50 and his relief was initially unaware of the treadle operation because events occurred before the handover could be completed. There is only one bearing failure every five million wagon miles, and only a very few result in a derailment — and only a few of the derailments result in a ruptured tank. It is virtually unheard of for a recently overhauled wagon to be involved in such an incident; the wagon had run only 640 miles since being out-shopped. And the tunnel represented only 1.5 per cent of the total journey. If it had been a story in a book, it would have stretched credibility.

Realising the train was on fire, the three-man crew ran the mile to the southern tunnel portal and used the phone at the signal there which connected to Preston Power Box rather than the local signal box to alert the signaller of the fire. The signaller alerted Greater Manchester Fire Brigade, who alerted their West Yorkshire colleagues. Using a pre-planned procedure, both brigades entered the tunnel from their respective ends, and both found small fires. The West Yorkshire brigade dealt with their fires quicker than their Greater Manchester colleagues and were leaving the scene, believing the fire was out, when one of the pressure relief valves on a wagon vented, putting flammable petrol vapour into the atmosphere. This was ignited by a still burning fire at the other end, and sent flames roaring over their heads (see photo 54).

They beat a hasty retreat and reformed outside the tunnel to discuss their next move. Meanwhile the Greater Manchester team persuaded the train crew to re-enter the tunnel to draw out the locomotive and the three non-derailed tankers, in order to remove 200 tons of petrol from the vicinity and to allow firefighters better access to the burning wagons. Once the train was out, the fire was clearly developing, and the fire crew also withdrew. The brigades could not communicate with each other directly by radio because of the terrain, and the police had to set up a relay station high up above the tunnel.

Hearing of the fire on radio news broadcasts on my way into work, I offered Scientific Services support to Douglas Power, the Chief Mechanical Engineer of the London Midland Region, whom I'd first met three weeks before following a fatal train fire investigation at Eccles. He accepted the offer, expecting — as we all did — that the investigation would be carried out the same day. Some hope! Arriving at the northern end of the tunnel, where West Yorkshire Brigade had set up a mobile headquarters, I reported to Derek Howarth, the Assistant Chief Fire Officer, who had overall responsibility of the whole operation. Luckily, he was grateful for rail-biased scientific support, but sensibly suggested there was nothing much to be done in the short term since the fire was well alight.

Most of the discussion centred around obtaining a 'best guess' of what would happen next. Ultimately it was decided to let the fire burn itself out rather than risk re-entering the tunnel, but that the ends should be plugged with foam to limit the air available to the fire, and thus slow down the burning process to minimise damage. Ultimately, about half a mile of foam was pumped into the tunnel, more than anyone had previous experience of. So, the question arose: what is the life of the foam, and will it prevent entry for weeks to come? In reality it decayed quickly.

The fire finally self-extinguished on 24th December but the tunnel was not accessed until the first week of January. Scientific Services staff carefully recorded the position and condition of each wagon, noting the condition of brickwork and rails in the damaged area. As well as massive distortion to the wagons, several of the bricks (mainly modern ones used for patch repairs over the years) had melted and run down the walls, and some of the rails had also melted (see photo 55). Samples taken back to the laboratory allowed the melting point of the brick to be determined at about 1240°C while the melting point of steel is about 1530. Around the top of the ventilation shafts, high up on the moors, hundreds of small metal spheres were discovered, resulting from molten metal being swept up the shafts by the hot gases generated in the fire, rising to a height above the vent and then falling to earth while cooling — a reverse version of a shot tower. Because of the severity and nature of the fire (could the tunnel have exploded?), the formal investigation was carried out by the Fire and Explosive Laboratory of the Health and Safety Executive.

Large parts of our submission to the British Rail internal inquiry were incorporated into the final Department of Transport report. Subsequently, I sat on the internal inquiry panel and among the questions put to the driver was what he was thinking as he was thinking as he ran out of the tunnel. "If Seb Coe can run a mile in four minutes, I reckon I can do it in three," he replied. When asked what his thoughts were on being asked to re-enter the tunnel to drive his train out, he said: "I was wondering whether my life insurance would cover being blown up after going back into a tunnel on fire." A railwayman's humour at its best.

On 18th November 1987, a fire broke out on the London Underground and had massive consequences for British Rail. To briefly explain: a match was dropped on a wooden escalator at King's Cross St. Pancras underground station. The match was taken by the revolving mechanism to the underside of the steps, where it ignited accumulated dust and grease. The resulting blaze killed 31 people and changed the understanding of how fire travels.

Prior to this tragedy, the perception was that flames always went up. But extensive computer modelling at the Atomic Energy Research Establishment pointed to the fire travelling along the escalator as if in a gully with sticky sides. Such movement was initially treated with sufficient scepticism to prompt the suggestion that the resulting diagram was being held upside down! A one-third scale replica of the escalator was built at the Health and Safety Executive's fire and explosive laboratory and confirmed the findings. The phenomenon became known as the 'Trench Effect'.

At the time of the investigation, British Rail had wooden escalators at several stations and an immediate request was made to examine their mechanism for dust and grease contamination. This task fell to Scientific Services. We established that the dust was mainly composed of flakes of human skin and fibres from clothing. The lubricating grease was found to have built up over decades; there was little evidence of cleaning having ever been done. The resulting gunge was easy to ignite (the fibres acting as a wick for the grease), so an immediate regime of cleaning was introduced with a longer-term aim to replace all wooden escalators with steel. We also looked at the composition of the paint used in underground locations since that employed at King's Cross had initially been blamed for the rapid fire spread, and for contributing hydrogen cyanide to the emitted gases.

A public inquiry was held under the chairmanship of Desmond Fennell QC, which lead directly to the Sub-surface Railway Station Regulations, restricting the use of certain materials in such locations and banning smoking. Perhaps bizarrely, Birmingham New Street Station was defined as being sub-surface, since there was a concrete raft above it supporting the shopping centre.

An additional consequence of the fire was that British Rail appointed a fire safety adviser: none other than Derek Howarth, who was in charge of the Summit Tunnel fire response when employed by West Yorkshire Fire Service. This was a bonus for Scientific Services since he and I had met under 'operational conditions', and he had seen our capabilities in action. Also, he was a great believer in science. We worked very closely for several years.

It became obvious that to incorporate passive systems into coach design — i.e. things which reduce fire risk rather than how materials behave after a fire has started, we needed information about what people actually do in a fire situation, so we decided to carry out evacuation trials. Health and safety legislation prevented us from setting fire to a train with people in it (even volunteers), so it was agreed to offer a financial incentive to encourage the 'passengers' to attempt to evacuate a stationary train as quickly as possible on the sound of a whistle. But who should we use as passengers? After much deliberation we decided that police cadets were likely candidates: they were fit, bright and being trained to react calmly in any situation and to use their initiative. At this stage, the BTP did not have cadets but we secured the services of 25 Metropolitan and 25 Staffordshire police cadets, and the financial incentive was a sum of money to their respective force's benevolent funds.

We acquired a redundant but very modern coach from the class 151 diesel multiple unit, which had been tested alongside the class 150 Sprinter to gauge which would be the best buy for BR. The 150 won and the 151 was quietly shunted into a siding and forgotten about. We set up cameras inside and outside the coach. The first team to get all 25 cadets out would get the money. There were various refinements, such as the generation of family groups, where the 'children' had to get out before their 'parents' but we were interested to see if the cadets would collaborate to get out, or would it be every man or woman for themselves? We did various runs, applied statistics to the findings and hopefully moved the concept of fire safety forward. Yes, we knew the situation was far from realistic, but we needed to have reproducible results so the consequences of any changes we made could be assessed.

The results were widely publicised and became the basis of one of the early computerised models used for fire evacuation studies now regularly applied to the design of large public areas. During the trials most of the tables in the coach were destroyed as the cadets used them as stepping stones in their attempt to get out first, but the most telling point was when one of the Staffordshire cadets was seen by her training sergeant (a man far more fearsome than a company Sergeant Major!) knocking down a Met cadet by the doorway. Taking her to one side, he said: "If you want to be a police officer you will need to learn restraint. Good police officers do not go around hitting people who obstruct them." I wonder where she is now.

The Stansted rail link, which approaches the airport through a tunnel, opened in 1991. One area of interest to the commissioning team was tunnel safety in the event of a train fire. The tunnel was the first to be fitted with a door-level walkway throughout its length so evacuation from a burning train could, in theory, be achieved rapidly and safely by simply stepping from the door onto the walkway — rather than having to jump, or more likely be pushed, onto the track below. The powers that be decided to carry out evacuation trials when the tunnel was completed but before it became operational. They procured the rolling stock and enough guinea-pig passengers to fill an entire train and arranged a date. I was invited to attend as an observer, together with several representatives of the police and other emergency services.

Unlike our trials, their passengers were from the Actors' Union and various local organisations, and they would be paid to be there, not paid by results. The train arrived in the tunnel, when passengers were briefed to evacuate it on the word 'go' and make their way along the walkway to the fresh air. However, the sense of urgency was missing, so the pace was leisurely. It was noticeable that many of the actors were treating the whole exercise as an opportunity to meet long-lost colleagues: it was more important to find out what productions they'd been in since they last met than to get to the tunnel mouth. A nice train ride, a pleasant gossip and a stroll in the country, all paid for — what could be better? I am not sure how the results were interpreted. The whole process was so completely different from ours that there would be little comparative significance. I still do not know whose approach was closer to reality.

CHAPTER 8

Oils and Condition Monitoring

Probably the most important role for Scientific Services, following the widespread introduction of diesel traction, was ensuring that engines were in tip-top condition by analysis of the lubricating oil. Our experience proved invaluable when the early High Speed Trains (HSTs) began to fail in service. We like to think we saved the iconic trains from being condemned as another BR fiasco.

ALL RAILWAYS consume vast quantities of petroleum products and British Rail was no different, writes Ian McEwen. From light oils for the lubrication of the thousands of railway clocks to lubricants for bridge bearings — and everything in between, not to mention diesel fuel — all were purchased by the Director of Supply's Department (DoS) with technical support from Scientific Services. (Although, before it became uneconomical the Derby Carriage and Wagon works produced their own greases, at one stage manufacturing seven tons a week).

For the high-volume items on the lubricants catalogue, the DoS purchased products not by trade name nor to BSI specifications but to BR's own specifications, drawn up and developed by Scientific Services chemists, with Lubrication and Wear Unit staff taking the lead role. Although quality control of this wide range of products was not carried out as a routine, any problems arising would undoubtedly result in analytical work in one of the laboratories. The oil bench was a key part of the laboratory in Derby, and most newcomers to Scientific Services would usually spend some time at this facility as part of their training.

Oil analysis

John Sheldon was in charge of the oil bench at Derby, and here he outlines the early days of condition monitoring: predicting potential problems in an engine by looking at the condition of the lubricating oil.

"Oh dear! You are all worn out. You had better see a chemist."

As steam locomotives were superseded by diesel power, the volume of oil testing and analysis increased dramatically as the era of preventive maintenance of diesel power units from data provided by lubricating oil analysis was born. For this to be possible, the Department of Mechanical & Electrical Engineering (DM&EE) had to be on board. The basic concept was to monitor the lubricating oil condition at regular intervals. The crankcase oil is the life blood of a diesel engine and, like human blood, much can be learned from analysing it. The oil analysis could provide information on two quite different fronts. Firstly, simple routine tests for the physical properties of the oil indicated whether it was fit for further service. Suitability for continued use of the oil as a lubricant was confirmed by the viscosity (too thin indicated fuel contamination and too thick indicated more than the specified quantity of 'dirt', i.e.

carbonaceous particulates). The latter was determined by a colour coding technique. Acidity and the presence of water were also determined: if water was present it would suggest a coolant leak. Secondly, since mechanical wear is inherent in all diesel engines and can be detected by the presence of metallic particles suspended in the lubricating oil, analysis to quantify such metals could indicate potential problems, especially if it showed a sudden increase between consecutive samples from the same engine. In extreme cases, it could suggest an imminent failure of a critical component. The unit in question would then be taken out of service before its scheduled examination to be checked out by depot engineers. Remedial work, if found to be necessary, would eliminate major expense and disruption if the engine had failed in service.

The spectrometer, the laboratory tool able to identify and quantify metallic elements in the oil, was a sophisticated piece of instrumentation in those days. After examining the success of the monitoring process, the DM&EE gave its approval that at every B exam (a scheduled depot examination by depot staff akin to a mileage-based service of a car), an oil sample would be taken and dispatched to the laboratory in an appropriately labelled 250ml plastic bottle. To enable ease of sampling, each power unit was fitted with a sampling cock at an appropriate position within the oil line.

Each locomotive's power unit had its own unique health card and once recorded, adverse results were transmitted to the DM&EE to take appropriate action: an oil change if required and a thorough engine examination if the spectrographic results warranted so. To enable the major maintenance depots to ascertain fitness for purpose of each locomotive's engine lubricant, they were supplied with basic viscosity testing equipment which indicated if the oil met acceptable parameters. This enabled the depot to perform an oil change at the scheduled basic A exam if the sample failed the viscosity check, rather than wait for the laboratory results from the B exam. Senior rolling stock inspectors were chosen for this task and were trained to use the rolling sphere viscometer. Laboratory staff made regular depot visits to ensure the accuracy of the equipment and check that depot personnel knew what they were doing.

Here Ian McEwen takes up the story. The pioneering work on spectrometric analysis of oils from BR's early diesel fleet soon showed clear benefits in helping identify engine faults and led to establishing condition monitoring routines. Modern crankcase lubricants offer long-lasting protection from wear and component failure. The monitoring programme allowed a 'change as required' rule for all large diesel engines' oils; that is to say, the oil is not changed on a pre-arranged fixed mileage or time basis but only when the laboratory indications are that the oil is unsatisfactory through contamination or general degradation. It was calculated that for engines with sump capacities greater than 90 litres, the laboratory costs could be offset by the savings from unnecessary oil changes. Mainline locomotives have sumps containing between 300 and 700 litres. Even for the smaller engines used in diesel multiple units (DMUs), which received fixed-period oil changes, there were advantages in monitoring the sump oil in order to highlight engine faults.

As John has already indicated, physical testing of the oils arriving in the laboratory was restricted to four properties: viscosity; total insoluble matter to indicate dirt content (see photos 57 and 58); water content to check for coolant leaks; and total base number (or acidity) as a measure of additive depletion. Fuel dilution from faulty injectors and coolant leaks were the most common reasons for recommending an oil change, fuel in particular lowering the viscosity of the oil and significantly affecting its lubricating properties. Because of the tight specifications drawn up for the oils at the time of their delivery, it was rare for an oil change to be needed due to its lubricating properties having deteriorated, despite the considerable distances travelled. This was also partly because at each minor service the oil level was topped up. It was calculated that on average, an oil change for a large locomotive was needed every 180,000 km of running. Because of the very tight controls, with frequent sampling, of HST power cars, an oil change every three years was the average — the equivalent to approximately one million km of travel. On a fixed-mileage oil change basis, these changes would need to be far more frequent. It was estimated that savings in excess of £1-million a year were made from the change on condition rule alone.

Apart from major overhauls, railway vehicles are serviced on depots overnight and this is generally when oil samples are taken. Occasionally an operative, keen to get the night shift over with, would take more than one sample from an engine and label the sample bottles differently as if coming from, say, the other engines in a diesel multiple unit. Such casual sampling would almost always be spotted immediately from the laboratory analysis. This would be reported in the results and any action needed left to the owning depot.

The HST effect

Here Ian Cotter and Ian McEwen contribute to the story of the HST, and how close scrutiny of its engine oil made it the saviour of British Rail (see photo 61).

In parallel with engineers in the Research Department developing the APT, the DM&EE decided to take the conventional train design and operation to its limit, just in case it was needed. They combined the concept of the locomotive hauled express passenger train with the diesel multiple unit, effectively producing an express train with a locomotive at each end, working in tandem so there was plenty of power, and with the operating freedom that having a driving position at each end allowed. The train would be streamlined but the iconic nose shape, designed by Sir Kenneth Grange and made possible by the use of plastics, was not present on the squat-nosed prototype. It appeared on all the subsequent vehicles. The engines were body mounted and coaches fitted with skirts to aid aerodynamic flow. The only flaw was the choice of engine: to speed the design process, an off-the-shelf engine was necessary, and the choice fell on the Paxman Valenta marine engine. But the DM&EE had a train ready to plug the gap unfilled by the APT, and one which was sufficiently different to let potential passengers know the railway was on their side.

The initial intention was to use a 16-cylinder version of the Valenta, however, constraints set out by the permanent way engineers in charge of the track regarding limits on axle load and un-sprung mass led to a compromise that a lighter 12-cylinder version of the Valenta, rated at 2,500 horse power, would be used and still be able to produce 125mph running. The Valenta was designed as a marine engine running at full throttle for hours on end. Its loading cycle in the HST was much more arduous and the engines suffered from unreliability in the early days of service as the stresses of thermal cycling and the continual stop-start operation in the rail environment took its toll. Engine components started to deteriorate rapidly. Major problems were encountered with the pistons, where the gudgeon pins that secure them to the connecting rod were failing. This resulted in diesel fuel leakage into the engine oil, making it thinner and hence less able to lubricate effectively. The piston could become detached from the connecting rod, causing it to flail about and damage the crankcase, releasing the oil from the engine in a black swathe down the side of the train. In the extreme case, the con rod could be forced through the side of the crankcase leading to a total engine failure. This situation was colloquially known as a 'leg out of bed'.

Capacity for repairs soon became strained and spare engines to replace failed ones were in short supply. With power cars sitting in sidings and trains cancelled, the engineers asked for help. More frequent condition monitoring with faster turnaround in the laboratory was the solution. A turnaround of maybe a week was acceptable for most of the BR fleet using heavy medium speed engines where defects took time to develop but was not appropriate when HST engines were developing serious defects every few days. The analysis system at the London lab, serving the Western Region where the first HST services were introduced, was swiftly overhauled, new technology employed and a service that was able to analyse samples in a few hours was born. Samples were taken from power car engines every two to three days and results were reported within three to four hours of receipt. Use of the HSTs themselves to transport the samples to London allowed oil taken on the night shift at Penzance to be reported by 3pm the same day, in time for remedial work to be programmed in by maintenance controllers. As things got worse the process became even more frenetic. Our phone call to a controller would elicit the question: "How long will it last?" If the answer was "maybe an hour or two", after a pause to check where the unit was in its journey, it might be swapped for another which had been due to go to a depot. Or, if our response was a more strident "stop and examine", the train could be stopped at a signal and the driver instructed to shut down the offending engine. This happened with a Penzance-bound train south of Taunton, with the controller saying he would have to find a loco to tow it over the South Devon Banks where running on one power car was prohibited. Thus, the oil analysis saved the HST, which itself was known as the saviour of BR. All Paxman Valenta engines have now been replaced.

Initially all laboratories except Swindon had input into condition monitoring but, as the analytical facilities became more sophisticated to handle the large sample throughput, the service was centralised at London and Doncaster, where HST samples from the Western and Eastern Regions were sent. True centralisation was eventually achieved when all condition monitoring was located to Doncaster. This involved complicated logistics to collect samples from a great many service depots and deliver them to the laboratory as quickly as possible. The laboratory's target was to test and report by 4pm on the day of sample receipt. Results were checked against pre-set rejection limits which called for immediate oil change and against trends from the last three samples from the same engine. They were then presented to the engineers responsible for the various engines via the system-wide RAVERS (**Ra**il **Ve**hicle **R**ecord**S**) database. The centralisation allowed investment in the development of advanced computer networking and automatic analytical techniques such as Inductively Coupled Plasma spectrometry (ICP), where a very small amount of the oil sample taken for physical testing can be examined by ICP to quantify suspended wear particles, additive elements and common contaminants. The central oil analysis facility at Doncaster was equipped with an analyser working via a carousel feed system and auto-sampler to deliver a throughput of up to 120 samples per hour. The hub was a Laboratory Information Management System (LIMS) which communicated with the analytical instruments, collating results as they were produced and forming them into batch files to be passed to the main database. The latter was held on RAVERS, allowing access by headquarter business engineering groups, depots and laboratories alike. This mainframe computer was used to run checks against action guidelines and trends from previous results from the same engine, before splitting into separate files of normal and adverse results with suggested actions for the latter. The success of the scheme ultimately depended on the skill and knowledgeable interpretation of the laboratory technician in charge, often in dialogue with engineering staff, and the action he or she recommended. False alarms can be costly and destroy confidence in such a scheme. Prevention of a catastrophic failure by early detection was shown to give significant re-build savings, quite apart from the consequences of disruption to services (see for example photos 62 and 63).

A major re-build — necessary when an engine suffered damage to pistons, liners, crankcase, connecting rods and crankshaft — was estimated in the 1990s to cost in the region of £170,000. Prevention of three such failures a year would have paid for the totality of the analysis costs for the HST fleet. Figures for the year 1990 showed 72 problems requiring action were identified, 15 of which were associated with piston, main and big end bearings. The laboratory-recommended actions included Further Sample (FS), Exam Recommended (ER) and, particularly for the HST engines, Stop and Examine (SE). This latter action meant it was possible with a critically severe result to stop a set at a signal and shut down one of the two power cars for examination on return to depot, though this was rarely used.

To ensure the demand for analysis could be met, automation of the process was essential. But it cannot be achieved overnight, especially if the process involved is a specialised one. Dave Smith takes up the story of how automation was brought in at Doncaster.

There was always pressure on the number of staff employed in oil analysis, particularly so as the number of samples processed increased. This was the driving force behind automating as many as possible of the test methods used. A team from the Analytical Services Unit — consisting of Phil Jerrison, Neil Fearn and myself — was asked to find ways of streamlining the processes with a view to reducing staff levels from 20 to ten. Our main activity was persuading the industry that condition monitoring was the future, and that the standard methods approved by the Institute of Petroleum could be improved upon without compromising accuracy.

The early spectrographic analyser for wear metals present in the oil was based on looking at the density of light passing through a film of oil. This technique was superseded by electronic detectors which looked at each spectroscopic line generated as the oil was sparked between two carbon electrodes. One electrode was fixed and the other was a rotating disc whose lower edge dipped into a small bath into which a few millilitres of the oil being sampled was placed. As the disc rotated the oiled area came between the two electrodes and a high-tension electrical spark was induced. The light from the spark was dispersed using optical prisms into its component wavelengths, representing the wear metals present in the sample. The amount of metal present was determined by comparison with a calibration chart.

Having read the description of the process, you will gather it was a time-consuming process for each sample! It was improved by the introduction of Direct Reading Emission Spectrometers, which had computers (but with limited memory; this was the early 1970s) that could read the concentrations directly and print out the results. When HST samples started to be handled, the computer interface was updated, using more modern technology to allow direct input into the vehicle records stored on a large IBM mainframe (the previously mentioned RAVERS). But there were still local records kept on the **Lo**cal **Ve**hicles **R**ecord**S** (LOVERS) system, so Scientific Services could legitimately say it was supplying both LOVERS and RAVERS to the DM&EE.

But it was still too slow to get the turnaround required for the HST fleet, so, with centralisation at Doncaster came the introduction of Inductively Coupled Plasma spectrometry, using electromagnetic induction to spark the sample at several thousand degrees centigrade, but again we were hampered by the lack of an automatic feed. We approached the manufacturer, ARL of Switzerland, and persuaded them to develop an automated system for their equipment to meet our requirements (see photo 60).

Every oil received had viscosity measured with a calibrated viscometer. The oil was not new and transparent but rather used and black, meaning it could be difficult to determine its passage through the viscometer with accuracy, although this was what was called for in the approved Institute of Petroleum standard method. We had banks of fish tanks with water maintained at the specified temperature, with staff timing the rate of flow of oil through up to six viscometer tubes submerged in the tanks, at the same time using a separate stopwatch for each tube (see photo 56). After every test the viscometers needed a lot of cleaning and had to be re-calibrated at regular intervals. The tests for carbon/soot and water content was equally time-consuming. We worked with a company called Trivector as they developed an automated system for viscosity and dirt using a PTFE tube in an aluminium block at a controlled temperature.

In this way the relative viscosity and dirt could be determined against standard samples by passing a slug of oil past two light cells. The accuracy with respect to repeatability was much improved and a carousel allowed an automatic feed. We recognised there was no need for absolute results for this work, nor the use of IP methods. Another project was initiated at De Montfort University, run by Professor Malcolm Fox and other researchers, who developed a process for automating the process of determining acidity. Again, the method was not approved by IP or ASTMS (the US standards body). The results showed better reproducibility than the standard approved method and could be automated.

The introduction of the automated systems (see photo 59) brought many benefits. The elimination of those tedious jobs that could be automated resulted in a reduction in manpower and cost. The IT systems had an artificial intelligence element which analysed trends. The system only highlighted those results where the variation was greater than typically expected, thereby reducing manpower involved in interpretation and allowing a focus on failing engines.

The operational benefits were realised and the cost reduction in testing meant it was then possible to extend the condition monitoring service to smaller engines. There was a significant increase in our automated systems and our success was recognised worldwide. This resulted in visits from oil companies and industrial organisations with condition monitoring systems in place and were attracted to the benefits of automation. Another very welcome outcome was that we received royalties on the equipment bought by companies setting up systems using the equipment we developed.

Ian McEwen takes up the story again, highlighting some of the unexpected consequences of condition monitoring. Although the sampling and analysis of HST oils became a routine exercise, it could occasionally highlight most unexpected trends and unexpected behaviour. One incident in the mid-1990s began when Doncaster laboratory reported higher than normal silicon levels on several engines. Rising levels of silicon can be an indicator of problems with the torsional vibration dampers on the crankshaft and could lead to significant damage if not addressed.

We were suspicious that several engines were involved, and subsequent samples showed a further rising trend. The evidence was pointing to an oil-related problem rather than engine. The normal silicon levels of 11-12 parts per million were rising as the engines were topped with fresh oil: a steady rather than alarming increase. While samples of the fresh oil from the depots where the engines were serviced were requested, I spoke to the crankcase oil supplier, but they were unable to explain what might be wrong. The additive pack used in the oil contains a cocktail of chemicals, each with a specific purpose and, as further information was gathered, I suggested to the supplier that the anti-foam agent (a silicon compound), was overdosed. I even suggested it may have been overdosed by a factor of ten.

This was laughed at by the supplier, who put it to me that the oil manufacturer and supplier, as well as their additive package supplier, had strict quality systems and such a thing could not occur. When, in consultation with the Director of Supply's lubricant buyer, we temporarily stopped the supply of oil and went to our second manufacturer, we were suddenly taken seriously.

In subsequent days it was found, and agreed by all, that the silicon compound was indeed at ten times its specified level. Neither of the two quality systems had identified this mistake and it needed our own laboratories to raise the matter in the first place. Our condition monitoring regime was put out of kilter for some weeks before levels returned to those expected and our engines were fully protected. It was left to the Director of Supply to sort compensation from the supplier.

And now for our final topic of this chapter. Locomotive traction motor gearboxes were notoriously difficult to seal and for a great many years were lubricated with residual lubricants. These were manufactured from the residues left when the lighter cuts of fuel and oil were taken from the distillation of crude oil, their main feature being that they were extremely thick and would thus minimise leakage. This, and the fact that tackiness additives were added, made them very difficult to handle. Additionally, they contained carcinogens and needed to be handled with care. We worked with lubricant manufacturers to develop multi-grade greases, which would be much safer and easier to handle.

At one point the lubricant was sold in polythene 'sausages' for clean-hands filling, the plastic getting mashed up in the gearbox and cleaned out at major overhaul. Various other methods were adopted at depots: warming the lubricant before use (particularly when supplied in 'mastic' type tubes), forming into cricket ball sized spheres and wrapping in clingfilm or newspaper. Again, the wrapping was cleared out at overhaul. The most creative means we heard about came from a railway depot in a developing country where a young lad was given this nasty job of topping up gearboxes. He formed the thick lubricants into balls and rolled each ball in clean sand to make them easier to handle and store. The balls were dropped into the gearboxes, sand and all!

CHAPTER 9

Miscellaneous

We have covered a lot of the work of Scientific Services from the 1950s. We were always up for a challenge, as the following stories — from pest control to the chairman's loo, by way of ghost trains — illustrate. Ian Cotter introduces the subjects he has selected which do not fit anywhere else in this book.

S CIENTIFIC SERVICES was asked to examine all sorts of problems and I had a fascinating career (most of the time!) looking at a variety of issues. As the railways became self-sufficient in manu-facturing everything needed for new locomotives and rolling stock and their operation, the chemist's job became more varied and metamor-phosed with a broader remit into the scientist. As the years went by the scientist was consulted on many matters and addressed many issues in the mainstream of the rail industry such as managing storage of fuel (see photo 65) but also on what might be termed the periphery — or just plain weird.

Pest control

The railway, writes John Sheldon, is a vast landowner and maintains thousands of buildings, let alone millions of miles of cabling. In town or country, one thing most passengers are unaware of is the constant battle against pests, be it the pigeon decorating the commuter with gay abandon, our friend the grain weevil or the rat chewing away at the signal cable. Pest control is both an art form and a science; a crude battle to the death or an intellectual exercise to design the rodent-proof equipment box. The railway employed both the rat-catcher and the eminent scientist to contain the potential for damage.

The railway industry's formalisation of pest control emerged in the 1940s and led by Horace Hayhurst, a leading entomologist of the day who headed the Infestation Section. Insect control was necessary to protect the myriad of natural products stored in granaries and ware-houses prior to or during transit. Horace wrote a book entitled Insect

Pests In Stored Products, with a foreword by no less a luminary than Sir Harold Hartley, vice president of the LMS and a leading advocate of railway research at the time, who lent his name to the building housing Derby Area Laboratory on its transfer from Calvert Street. He was, perhaps appropriately, also Controller of Chemical Warfare at the end of the First World War.

During the 1960s, the section was managed by David Jenkins who left in the early 1970s to join the private sector and subsequently form his own company. Jenkins, who changed the name to Disinfestation Section to indicate that the idea was to eradicate the pests rather than encourage them, was succeeded by Alan Waterhouse, from Crewe laboratory. By then the emphasis was changing from stored products to general pest control, focussing on BT Hotels, Travellers Fare, Sealink and Seaspeed. Most rodent control was undertaken by contractors hired by BR's civil engineers. The in-house rat catchers were mostly redundant by the early 1970s. At this time, the Disinfestation Section as it was still known employed 12 pest control operatives divided into four teams of three based in Manchester who travelled to all parts of the BR network at the start of each week. Subsequently this modus operandi was considered an inefficient and expensive enterprise without a structured plan.

In 1976, environmental health departments, particularly in London, were concerned that BR's piecemeal approach, especially in complex railway termini, gave a poor level of pest control and were seeking a 'block control' policy. This led to a rethink of how the service worked and how greater efficiency may be achieved. The Manchester operation came under the spotlight and in 1977 its role was expanded. The section was renamed the Pest Control Unit and additional resources were created in London, managed by Richard Strand and with a couple of posts in Glasgow under the management of the area laboratory.

With the sale of so many of BR's subsidiaries in the early 1980s and the need for greater cost control, the Pest Control Unit was wound up in 1985 and the task of privatising pest control work to become an effective, organised service provided at a realistic price, fell to Scientific Services' newly created Derby-based Products and Services Unit (PSU), headed by Richard Strand. As its name implies, the PSU was responsible for more than pest control, and with the introduction of the Quality Assurance standard BS 5750 (ISO 9000) it provided lead assessors and assessors to inspect the QA systems of BR's suppliers in areas where scientific input was necessary.

With invaluable assistance provided by Westminster City Council's chief environmental health officer, documents were drawn up for tenders to be offered to the private sector for pest control provision for each of the BR regions. Prior to being invited to tender for one or more contracts, pest control companies were vetted by PSU staff. Once up and running, contractors' performance was monitored, and quarterly progress meetings were held with each region. These meetings were chaired initially by Richard and included the contractor (who traditionally bought a post-meeting pub lunch for attendees), Travellers Fare's technical manager and the regional representative, usually a clerical position. Richard subsequently left the PSU and BR to become executive director of the British Pest Control Association (BPCA) and was succeeded by me. I was given the title of BR Pest Control Contracts Manager.

The contract specified the control of rodents, cockroaches and other pestilential insects. Feral pigeons were excluded but nonetheless provided a substantial income for the contractor. Regular culls occurred overnight as birds roosted to reduce the population of these rats with wings, as one contractor described them. Pigeon proofing to protect buildings from their deposits was also quite remunerative. They were exterminated by contract staff armed with high-powered air rifles and assisted by a floodlight. Despite having a strong beam pointed in their direction, the birds never moved and were sitting ducks (or should that be pigeons). On one occasion as our sharpshooting contractor was about to commence this work on the roof of Euston station, he was spotted by a member of the public who rang the police. The constabulary turned up mob-handed and fortunately our man was able to produce his bona fides, and no harm was done — unless you were a pigeon. The fault lay with the station personnel whose duty it was to inform the police of what was taking place on that night.

Everything the contractor did — from planned routine inspections to call-outs — was recorded in logbooks sited at major locations, where he would detail his findings and resulting actions. Each page was triplicated: one copy to be retained, one for the contractor and one for the PSU. One such report concerned the sighting of a white rat on the concourse at Cannon Street station. No extermination was required since the sighting was no rat but a hamster that had briefly escaped the clutches of its young owner.

Each contract has a clause stating that should the contractor fail to meet its terms, he would be given due warning to meet obligations and, failing that, the contract would be terminated. It's never good to have to terminate a contract but I found myself in this unenviable position with a major player in the pest control business who held the LMR contract. Birmingham New Street station had a major mouse infestation and, despite having ample opportunity to rectify the situation, improvements were not forthcoming. Clive Wadey, Travellers Fare's technical manager, was equally concerned, so as a short-term measure, I invited the Western Region contractor to take over since his past performance was faultless. Meanwhile the incumbent contractor's point of contact, the technical director, was informed of the termination intention verbally and formally in writing. Needless to say, he was not a happy bunny and invited John and Clive to lunch at Brown's Hotel in Belgravia, ostensibly to discuss the matter further. It was a first-class meal that finished with port and brandy and a terminated contract. The Western Region's contractor saw out the remainder of the contract, having successfully rid New Street station of its mice and was subsequently awarded the LMR contract when it came up for renewal.

Pest control throughout history has attracted many oddball suggestions to enhance or replace traditional methods. Ultrasound to deter rodents and children's toy windmills placed in the ground above a mole run are just two ineffective ideas. A letter appeared in my in-tray from a whizz kid in the City who claimed to have the ultimate answer to pigeon deterrence. He theorised that birds use the earth's magnetic field for navigation, so that if bar magnets were placed in sensitive areas, pigeons would be confused and move on. I met our budding entrepreneur in his office in Paternoster Square and agreed to a field trial.

"Nice of Mr Sheldon to use magnets to direct us to the peanuts!"

With Clive Wadey's cooperation, an island food kiosk within Victoria station's concourse was the chosen test site, a generous portion of peanuts scattered on the roof. Bar magnets, magnanimously provided by the instigator, were strategically placed to protect the ware. Clive and I stood back to witness a small flock of pigeons descend on the kiosk and gorge themselves on the unexpected nutty cornucopia. The city boy was thanked for his interest and feral pigeon control continued by tried and tested means. Back at PSU, I placed the inherited magnets on my filing cabinet and moved on. Some weeks later, retired after almost 40 years' service as a BR scientist and at my leaving do, Ian McEwen recounted the magnet experience. He dryly remarked that while the magnets had no effect on pigeons, Gordon Anderson, an office colleague, twice got lost on his way home.

You just can't tell some people

BR Research had many distinguished visitors, writes Dave Smith, and those from government were regarded as a necessary evil as we approached privatisation. We were supposed to be more productive, but these visits were disruptive. I, for some unknown reason, was asked to look after a senior official from the Ministry of Transport and let him see the valuable work Scientific Services undertook. The directors explained they felt that, as he was a politician and not a scientist or engineer, we need only show him some of the less complex areas of work. They thought (rightly, as you will see) that the high-tech work undertaken in many parts of BR Research would have been too challenging for him comprehend. A few days before the visit, the message I got from the Gods made it absolutely clear that the visitor was a technical lightweight.

The day did not start well when I introduced him to John Harding, who was leading on environmental management issues and the production of BR's Green Book. John asked of his view on some aspect of policy and the visitor, in typical politician fashion, spoke a lot but said nothing. His final comment was: "You will have to wait until the green paper is produced." We moved on to the forensic room, where we'd laid out some of the cable used on the overhead electrification of the King's Cross to Edinburgh route. The catenary wire used is an expensive special copper alloy and was being stolen as it was being erected along the East Coast Mainline before the system was completed. I thought I was being witty when I commented: "We have a current problem with the theft of electrical cable!" My pun went unnoticed. I carried on, explaining our role was to examine the chemical composition of stolen cable to provide evidence in court. I explained that the theft problem was only temporary and was surprised when our visitor asked why. I explained that when the work was finished, the cable would be energised. The official asked: "Why would that make a difference?" I explained that energised meant it would have 25,000 volts passing through it. He was just asking what difference that would make when his assistant took him to one side. I decided to move on!

I then introduced him to Dick Hall, who was showing him the work being carried out on high visibility signs for use on the track by permanent way staff. It was apparent from his questions that he could not understand the problem with mixing men and trains, let alone a need to find a solution. If he had appeared as a character in the TV programme Yes Minister, he would have been considered unbelievable!

Things that go bang

Some analytical tests were carried out very infrequently, writes Dave Smith, and you had to devise your own method — unless you could track someone down who had written a methodology, rather like Mrs Beaton's cookbook.

I was unable to find a method for one such job. There were still some steam engines around and they used a thick black tannin solution (called boiler tan) as a corrosion inhibitor in boilers. We were required to test this to see if it met the specification of thick black goo (organic myrobalan), sodium nitrite inhibitor and water.

It was determining water content that nearly led to my early demise. On the oil bench in the main lab, they had some fancy glasswork with condensers and a flask on an electric heating mantle. A standard method for measuring water was to boil it up with xylene, a toxic flammable organic liquid which vaporised with the water in the condenser, allowing the heavier condensate (water) to collect in a graduated side arm. The method was used successfully on a daily basis for other purposes and normally worked well.

I had obviously forgotten my earlier schoolboy years, otherwise I might have thought twice about mixing an oxidising agent (sodium nitrite) with organic material (sounds a bit like a gunpowder mix). All went well until all the water had, as planned, evaporated. However, the gunpowder experiment then took over. Flames appeared in the flask and the equipment shot towards the roof with tubing and glassware going in all directions. The impact on the wall and ceiling mattered not, since the scars added to the character of the place which had seen so many memorable events. The event was short-lived and thankfully no one was killed or injured.

Having sort-of recovered and being surrounded by my much-amused colleagues, I said: "Has anyone done this analysis before?" Someone remembered that Granville Morley had analysed it a year or two previously, but he'd chosen to evaporate the water off in an oven and weigh the residue. I asked: "Oh, was that a better method then?" The reply: "No! The oven is over there — the one with the distorted top." It turned out Granville had adopted the oven approach after Jim Ward experienced the same disaster as I a few years earlier. Nobody seemed to bother about such events and it was accepted as part of the rich tapestry of life (or death!).

All at sea

The BR senior marine engineers had meetings at intervals to look at common issues with the HQ staff and the local port senior engineers, Ian Cotter recalls. The London laboratory acted as liaison with HQ as it was located in the capital; occasionally there was an issue that involved some laboratory work requiring a Scientific Services representative to attend the meetings. On one occasion the engineers decided to use their maritime facilities rather than meeting in the London offices. They chose the crossing from Harwich to Hook of Holland and back without landing, holding the meeting and dinner in one direction and occupying cabins on the overnight return after an evening in the bar. Thus, there was an all-expenses paid trip for the laboratory representative — another perk of working for BR Shipping! The vessel we travelled on, if I recall correctly, was the MV St. Edmund which shared the work with MV St. George. Later the St. Edmund was requisitioned by the Government to carry troops and vehicles to the Southern Atlantic during the Falkland War.

Electricity for chemists

The London laboratory served the whole of the then Southern Region, which was peculiar in having mainly third rail electrified lines on suburban routes (it was colloquially known as The Tramway by the other regions), writes Ian Cotter. One of the many issues we were asked to look at was the selection of materials for the insulating beam supporting the third rail electrical pick-up shoes on each train. These were usually made from laminated impregnated wood to provide stability and strength, as well as being (apparently) a good insulator. In the lab we took on many tasks of which we had no experience, but as good scientists we worked out a way of doing them.

The test required a 650-volt DC electrical supply to be connected to two probes a short distance apart, possibly half an inch, and then a contaminant solution (typically salt) applied drops-wise until a current passed between the probes, showing a breakdown of the insulation. The equipment was home-made. A Perspex box with a microswitch on the cover for safety was constructed, connections made to a variable transformer so we could get the 650 volts from the 240-volt public supply. A trainee chemist was given the job of dripping the solution until it tracked. Modern health and safety would be somewhat critical of this approach, but it never caused a problem and provided the Southern Region engineers with the data they needed!

Chemists cleaning trains

Most service industries use off-the-shelf cleaning agents to meet their needs, writes John Sheldon. In most instances BR was no different, however, it did have a specific cleaning problem that could not be addressed by proprietary materials: ensuring the continued cleanliness of the exterior of carriage rolling stock. The major problem was the removal of brake block dust which occurred when the cast iron brake blocks were applied to the steel wheel rims. The resultant iron oxide particulates could adhere to the surface of the coach's bodyside paint and rapidly rust, leaving an unsightly brown stain.

This was traditionally removed with a strong acid but with the introduction of metal-bodied vehicles a less reactive cleaning agent required to be developed. Brian Buckley, Derby's cleaning guru, later to become Area Chemist at Doncaster, solved the problem by recognising that the less aggressive oxalic acid (a poisonous organic compound) was able, when in a water-based solution, to dissolve the stubborn residue and thereby effect a clean carriage. Producers of cleaning materials were contacted with a view to supplying the vast quantities of oxalic acid required throughout British Rail. The product was to be called Exmover, the term coined by Brian.

Cleaning plants, fitted with rotating flails as now used on car washes, already existed at maintenance depots but had to be modified to accept the new cleaning regime. Being acidic and toxic, the effluent needed to be neutralised before discharge to sewer. This was achieved by dosing the run-off from the coaches with sodium carbonate. Cleaning plant operatives were issued with pH papers, the colour of which varies with acidity, to ensure the effluent was chemically neutral before discharge to sewer. Laboratory staff would pay routine visits to make sure plant staff were 'on the ball' in this respect and also check that concentrations of Exmover and neutralising agent were accurate (see photos 66 and 67).

In common with other activities carried out across all laboratories, regular progress meetings were held. Each area lab's cleaning specialist became part of the Cleaning Committee and labs would host meetings on a rotating basis. The committee was probably different from others insofar as it became a party piece for the chair, Brian Buckley. Brian was passionate about his pet subject and spoke incessantly throughout, leaving committee members somewhat nonplussed about what had transpired. Fortunately, Brian's deputy, Geoff Toole, acted as secretary and when he published the minutes, the confusion cleared somewhat.

Alternative methods for problem solving were often considered if they could be more efficient in terms of time and money or used less hazardous materials. Such was the case with carriage cleaning. Apart from the hazardous nature of oxalic acid, it was also expensive. Thinking outside the box, Eric Henley, Derby's Area Scientist at the time, and his cleaning specialist Andy Mailer came up with the notion of a sacrificial coating. A trial was undertaken by spraying a clean vehicle with transparent coating that would protect paintwork by preventing the adherence of brake block dust. Inevitably the dust was still created and stuck to the bodywork now protected by the coating. An alkaline solvent was then spray-applied to dissolve the coating, followed by a water rinse and reapplication of the coating as necessary. The result was a clean bodyside. Despite the apparent efficacy of the technique, it never became operational, British Rail preferring to stick with the tried and tested. With the introduction of new rolling stock incorporating disc and regenerative braking systems, the demand for Exmover as a carriage cleaning agent was gradually phased out.

Chemists cleaning cars

Railway coaches were not the only items affected by brake block dust. The cloud of hot iron filings which shrouded the wheels and bogies would distribute itself to anything close to the railway. The Beeching cuts of the 1960s resulted in the closure of many goods yards at those stations remaining open, writes Ian Cotter. BR, observing the increasing car ownership, started creating car parks next to stations in these redundant yards, allowing passengers to park and catch their train. The car parks were neatly positioned to catch this cloud of brake dust emitted from trains braking for the station and the complaints started to roll in about apparently irremovable brown stains. BR has long known about the brake dust issue on rolling stock and Scientific Services developed various cleaners to clean them at regular intervals. One was hydrochloric acid based, with a thickener so it stayed on coach sides to lengthen the reaction time and called Wundergunge, but the hazards of this are obvious. However, it was used for heavily contaminated vehicles.

The best compromise between safety and effectiveness was found to be oxalic acid (the basis of Exmover) a fairly mild (10kg to kill someone!) poisonous component of rhubarb leaves. Car park users were advised to obtain this from a pharmacy and use it to clean their cars, or we could post them some with instructions for use. Usually a local garage near affected stations became aware of the method and provided a service. However, some irate customers wrote to senior managers, prompting a memo from them to Scientific Services usually in the vein of: "Go and see him and sort out his problem." Thus, a chemist would be despatched with the equipment to wash a car and polish it up.

For the London laboratory, these trips were often to the Home Counties stockbroker belt where the owners of high value cars often provided a quotation from a garage for thousands of pounds for a complete respray. The owners were usually busy working in the City, so his wife or housekeeper would be at home to receive the expert car washer. Such visits involved using a van or car to get there; a method of transport rarely encountered by railway employees. In those days railway duty passes were the order of the day. I imagine in more modern times a chaperone might be needed! The car was usually cleaned satisfactorily and the claim averted, however, the story had it that some coach-built cars, like Rolls-Royces and Bentleys, could be damaged by the oxalic acid and would have cost BR a considerable sum in repairs.

The ghost train

In 1976 BR's Great Northern & City line took over the former London Underground tunnel from Drayton Park to Moorgate, writes Ian McEwen. With the consequent change of rolling stock, it was found that in some curves the bullhead rails wore down to their 'sidewear' limit in 10 weeks. Rails in tunnels cannot be turned to offer the opposite face because of lack of room and the wear patterns on the low rail made it difficult even to transpose the two rails. Fifty track-mounted flange lubricators were installed to alleviate the problem and I was involved in assessing the effect of these and predicting the extended wear life.

My first visit to the tunnel was with Jim Feeney, who had already made some rail profile measurements. As we approached the permanent way gang's hut just outside the tunnel, Jim predicted where each of the track team would be seated and what they would be doing — and he was absolutely right. On subsequent visits we would call in to say what we were doing, and it would be as if the men had not moved since our last visit. As we left the hut that night, one of the men said: "Look out for the ghost train!"

Our plan was to catch the last train into Moorgate and then wait for quite a long time on the platform until Control confirmed that the electrical supply to the third rail was switched off and it was safe to enter the tunnel. We then walked back the way we had come towards Drayton Park, stopping at various marked curves to take measurements. Jim was the one down on his knees with his nose at rail level while I was the one standing and recording his shouted readings for download to a computer later.

After an hour or so of getting used to the quiet of the tunnel and the total darkness, apart from our own lights, I heard a train in the distance rattling along and, from the increase in sound, coming towards us. There was also the air pressure pulse on the back of the neck as it got nearer.

Common sense told me it couldn't be on our section of track because the electricity was off, but my senses told me otherwise. Just as my panic was at its height, there was a big pressure pulse and whatever caused it was suddenly past us, growing more distant and the noise decreasing.

Was that the ghost train? No, of course not. It was the sound of traffic movement in an adjacent tunnel, ensuring that rolling stock was in its right place ready for the first service of the morning. It was, however, most alarming at the time. Later we heard another puzzling noise — the rail began singing slightly and a single white light approached us from the direction of the tunnel entrance. The light took quite a time to reach us and turned out to be the permanent way supervisor pushing a track trolley stacked with equipment.

Postscript: Work on underground railways obviously has to be done within a tight time window in the early hours of the morning when the tunnels are safe. The third rail is isolated and no traffic runs. We did work on Merseyrail in sections of tunnel that were very wet because of shifts in the water table. At the end of the night our challenge to each other after a tiring shift was to run up the down escalator at Lime Street before going back down again to collect our equipment.

The big red key

The original railway tunnels of the mid to late 1800s were cut by navvy gangs, working from each end and from intermediate points accessed by shafts from the surface down to tunnel level, writes Ian McEwen. The rough cut was then lined with brick or stone to give a uniform profile along the tunnel length, often leaving a void between the lining and the rock behind. As water tables have moved over the years many tunnels have become leaky, with groundwater in the void finding its way through lining joints to drip from the ceiling or run down the walls. This can affect the corrosion of components, such as rails and fastening, and also in some cases affect track circuits.

The Concrete and Building Materials Team contributed significantly to a large project with European funding aimed at waterproofing leaky tunnels. Permission was given to carry out a large-scale field trial in the disused bore of Alfreton Tunnel (the parallel bore is still active). A site was set up mid-tunnel in a wet section, with scaffolding reaching up to a platform under the roof from which foam could be injected under pressure into the void. This foam would then solidify and form a waterproof barrier. Laser beams and targets were set up to ensure the pressure did not distort the lining or bring it down.

On the day in question, as manager in charge of the team I was at the site to see the work being done. I climbed up to the platform and was having the process described to me when we were suddenly aware of blue flashing lights at the tunnel entrance 300m away. A fire engine approached — quite a surreal scene — and stopped at the foot of the scaffolding. Site manager Phil Ridgway climbed down to speak to the fireman in charge, and it soon became clear why they were there.

On the A38 road which passes over the tunnel, there is a lay-by. At the side of this is a round brick structure — the top of an air shaft for the tunnel. A rep having a break sitting in his car wound down his windows and heard shouting from the shaft which, at its lower end, was only a short distance from where we were working. The shouting was indistinct, but the driver thought someone had fallen down the shaft or may otherwise be in difficulty, so phoned the emergency services. A short time after this, the fire engine appeared at the tunnel mouth and drove in. Phil and the fire officer agreed that what was heard by the motorist was most likely our team shouting to be heard above the noise of generators. However, since our disused tunnel and the operational tunnel were connected at some points, the fire crew had to be satisfied with this explanation by fully inspecting both tunnels. This involved contacting control to stop all traffic from passing through the live tunnel until the all-clear was given.

Just before the fire engine left to back all the way out of the tunnel, Phil remarked that the huge doors at the tunnel mouth are usually closed and padlocked and asked if they had keys to railway padlocks. The officer in charge laughed, and pointing to the fire engine, said: "That's the only key we need. It'll push most things out of the way."

Making a splash and being a gentleman

In the late 1960s and 1970s, the Railway Technical Centre (RTC) was regarded as a centre of excellence for railway research and development, with visitors coming from all around the world to discuss and to learn writes Ian McEwen. With Scientific Services in Hartley House working to support the many departments of British Rail on a daily basis, a further group of chemists across the road on the main RTC site were carrying out long-term research on a variety of topics. The centre was very much akin to a university campus, with a good deal of freedom to think and to carry out original work. With both research departments and HQs for the engineering functions, the site had all the support services expected of such a campus. There was a very good technical library in its own purpose-designed building and a medical centre covering accidents in various workshops and even flu jabs in winter. The nursing sister was known to one and all as Chuck or Chucky, since this was how she greeted her customers. There was also an on-site branch of the Midland Bank, and an excellent canteen and officers' mess.

In line with the university approach to research projects, we took on young people in the process of gaining academic qualifications on a temporary basis. One such person was a young lady with excel-lent A Level results, taking a year off before going to Oxford University. To save her embarrassment, let's call her Judy. The fine canteen building was landscaped outside with shallow ponds that flowed under the glass front wall, such that a small area of the pond was inside the building. The main eating area was on the first floor and because of the popularity of the subsidised canteen, at busy times people queued up the stairs.

On the day in question Judy joined a group of us going to the canteen. The queue was so long we had to stand on the ground floor. As we chatted Judy forgot where she was and stepped backwards, only to lose her balance and fall into the indoor section of the fish pond (see photo 68). She ended up sitting with water up to her waist. It was a winter's day with ice on the outdoor pond surface, and the shock of the cold made her jump up, only for her to lose her footing again. She finished up almost lying down in the water. We showed great concern in helping her back to dry land but the unsympathetic crowd on the stairs cheered the event. With a good deal of aplomb, Judy gave a bow to the onlookers as we looked after her immediate welfare and sent her home. Within a week the pond had guard rails on its inside area.

Gifts of a trifling value

Ian McEwen salves his conscience here. I was working on a project aimed at predicting the wear of wheels and rails, particularly in curving situations. I received a letter from someone working on one of the heavy haul mining railways in Australia. The railway carries iron ore in huge freight wagons from the mines in the hills down to the coast, where ore is loaded onto ships. The letter's author was soon visiting the UK, familiar with our published work and requested a meeting at the RTC to discuss common interests.

Some weeks later we met one afternoon in my office. The next two hours resulted in a fruitful discussion for us both as we talked about rail and wheel steels, vehicle suspensions, wheel profiles, curvature of track and more. As we wound up the meeting, my visitor took a small black box out of his briefcase, obviously, I surmised, intended as a keepsake to mark our exchange of information.

He then told me that occasionally, along with the iron ore mined in huge amounts, they found very small seams containing opals, maybe once in many thousand tons of ore. These opals were of a colour not found anywhere else in the world. Because of their rarity and the fact that opals are the national gemstone of Australia, they were sold on to be used in high-end jewellery. I listened attentively, knowing we were not supposed to accept gifts other than those of trifling value — a diary or a calendar were the most usual. Eventually he slid the black box across to me with the words: "I hope you will accept this gift as a thank-you for your hospitality." I opened it expectantly to find a pair of cufflinks, each with a hemisphere of… highly polished dark grey haematite; that is to say, ferric oxide or iron ore. The cufflinks are in fact quite smart and look very good in a dress shirt with a dinner suit.

The moral dilemma as to whether I should have accepted this gift is clear. The cufflinks are now 30 years old and, of course, will have appreciated in value. Similar haematite jewellery can be found on the internet nowadays for about £10.

More pests

Roger Hughes describes the Invasion of the Maggots. As I was a fisherman, it was decided I would be the right person to investigate the safe carriage of maggots. They kept escaping from their containers and crawling into adjacent goods, carpets, boxes of vegetables and so on. I went to the Don Maggots farm at Mexborough to collect four gallons of maggots in large tins. The lid of each tin was pierced with small holes to enable moisture and carbon dioxide to escape and air to enter. I was going to determine the minimum size of hole that would enable the maggots to survive a 24-hour transit.

On my return in the late afternoon, I put the tins in the bomb calorimeter room adjacent to boss's office and went home. When I arrived the next morning, the girls were sitting on the bench tops saying: "You're in trouble." The maggots must have read The Great Escape: they were everywhere, rolling down the stairs like a living carpet, in the boss's office, even on the second shelf of his bookcase. I swept up as many as I could, and life returned to semi-normality for a week. Then bluebottles appeared, in swarms. Pest control was summoned from Crewe, who decided to use gas and extract to the outside using the lab's ventilation system. They turned it to 'blow' instead of 'suck', and we had to evacuate the building. The good news: I did eventually provide Don Maggots with an effective lid design.

Don't panic, we are all doomed

Dave Smith tells of the day the station master from Leicester called in a bit of a panic. He suggested the whole of Leicester station reeked of petrol. When the manhole covers were lifted, it was evident there was a significant amount of petrol in the drains. Since the station did not have any petrol supplies, the source was a mystery. Our immediate response was to enquire what action he had taken.

Recognising the potential for a build-up of fumes in the drains, he had ordered that all the manhole covers be removed and left off. However, this was at a time when smoking was permitted, and train drivers may well have thrown lighted matches or nub ends out of their cab window as they waited at the station. Additionally, train brakes used cast iron blocks on the steel wheel which caused sparks when applied, so were another potential source of ignition. At a distance of 30 miles it was difficult to judge the risk, but I advised him to replace the covers and use water hoses to flush as much petrol away as possible while we got a plan together. Peter Middleton was dispatched and the fire brigade called in to provide absorbent pads to help reduce the risk.

A nearby garage with its own storage tanks was responsible. A tank was leaking and the entire grassy bank on the east side of Leicester station was saturated with petrol. The garage was advised, and they took immediate action. We taped off the whole of the bank, put up warning signs and dealt with the petrol entering the drains.

In retrospect, and in today's safety conscious society, the station probably should have been closed and evacuated. What we did worked but it was a questionable decision. There can be a strong smell and evidence of oil/petrol on water, but the nose detects levels that may actually not be an explosive risk. At a distance it is not easy to judge if fumes are sufficient to be at the explosive limit, but on this occasion, I theorised this was very unlikely in the drains. The bank could have been a problem because a cigarette end could have ignited dead grass, for this to then cause fire to spread along the bank. What would our excuse have been if the station had blown up? Whenever I passed through Leicester station by train for many years after the event, I noticed the warning tape. No one challenged why it was there and whether it could be removed.

The chairman's clock

Finally, Vince Morris reveals why the chairman may have been even madder than us! This anecdote concerns one of our own staff, who perhaps got a bit carried away with the science.

The chairman of British Rail had well-appointed office accommodation, with en suite facilities for a wash and brush up before meeting passing dignitaries or cabinet ministers. In the time of this particular incident, the chairman had to relocate his office and asked that his (or was it BR's? — no matter) mercury compensated clock be moved to the new location. Such clocks use the expansion and contraction of the liquid metal so beloved of schoolboys to automatically maintain the pendulum length constant, so the accuracy of time keeping is not affected by the rise or fall in temperature. The clock was duly moved, but in the process some of the mercury spilled onto the floor of the office (see photo 64). It was subsequently swept up by the cleaners who, finding this strange liquid in their dustpan, decided to flush it away rather than risk an escape from the bin bag. It was thus placed in the pan and the loo was flushed.

"I wonder if all of them are as mad as the chairman?"

Having a high density, the mercury was not impressed by the flow of water intended to wash it into the sewage system and obstinately stayed at the bottom of the pan. Every time the chairman used the facilities, he was able to see a reflection of himself. He contacted the office manager, who contacted the scientists. Our colleague, probably the most able environmental chemist among us, was sent, armed with a pipette to suck the offending liquid out of the pan and dispose of it as a hazardous waste. This he successfully achieved in a few minutes. Then, the consummate scientist he was, he started thinking and penned a letter to the chairman explaining that exposure to mercury was well known to affect the mental capacity of humans. Therefore, all decisions taken by him during the previous fortnight and probably for the next fortnight should be ignored as being potentially unsound. I censored the letter before it could be sent... I sometimes wonder whether that was a wise decision?

The Authors

THE BOOK is the brainchild of Dave Smith, who wanted to write something down so his grandchildren would have some idea of what granddad did during his working life. He soon realised that the task of producing it was not for him alone: colleagues had similar stories to tell, so he invited several to contribute to this unofficial history of the Scientific Services division of British Rail Research. With more or less enthusiasm we agreed. But how to combine so many different tales and writing styles? The answer seemed to be to have discrete, but often overlapping, chapters each outlining an aspect of the work carried out by the railway chemist. Each chapter has been contributed by a 'champion' who has gathered the items concerning his expertise and edited it to ensure no repetition, verbosity or outright slander. With so many contributors, it became clear it would be difficult to establish a common writing style. Being pragmatic chemists, we know when we are beaten. So rather than trying to impose a style, we have identified the author of each anecdote and left their piece in their style. Of course, most of the stories — and all the contributors — predate the 'computer age', and since most of the official files have been spread to the winds, we have relied heavily on our memories. The authors started their railway careers in the 1950s, 1960s or 1970s, so forgive a combination of forgetfulness, exaggeration or confusion. Allowing for that, each anecdote is true: we know — we were there!

Children say they want to be a train driver, but in our experience no child has ever told their parents that they want to be a railway chemist when they grow up. So why did the six of us, and our colleagues over the past 150 years, finish up in what was one of the most interesting jobs around? Each author explains their drift into the sometimes surreal world of Scientific Services.

John Sheldon

Having decided that a university education was not for me, I left the cloisters of Belper's Herbert Strutt Grammar School aged 16 without a clear idea of what career direction I would take. The school's careers guidance was a waste of time. The less academic kids were advised to take up a trade or head to the shop floor of a local factory. The brighter ones were advised to stay on in the sixth form, with university the ultimate goal.

The likes of myself, a relatively bright kid with six O Levels under my belt who wished to leave, left the advisers in a bit of a quandary and their only advice was that I stay on.

After a lengthy summer holiday, it was time to make a decision. The school subject which I found to be of greatest interest was chemistry (one of my O Level attainments). With this in mind, I investigated the prospect of finding a job where my interest would also provide a remuneration. Dad was a railway signalman and had heard that Derby had laboratories within the compass of British Railways. So, having obtained a relevant address from the father of a friend of mine, a BR clerk, I wrote off and included my CV that was nothing more than my O Level tally. I also posted a similar application to Rolls Royce but on Dad's advice decided on the railway option, since he wisely pointed out that it wouldn't involve travel costs getting to work. My village, Whatstandwell, had — still has — a station, and railway employees were afforded free travel. Subsequently I was summoned to the area laboratory in Calvert Street, Derby, to be interviewed by the Assistant Area Chemist, Mr L G Tomlinson (Tommy). I must have impressed him, for shortly afterwards I was offered the post of technical assistant. I remember Tommy advising me that the most valuable part of the chemist's armoury was his or her nose, or sense of smell. On 11th October 1954 I crossed the threshold of the laboratory (previously Midland Railway's electric light power station, with a large chimney still dominating the vicinity) to start a career as a railway boffin that would take me through to retirement.

What made the Stinks work so interesting were the various disciplines I encountered. My involvement in lubrication technology is well documented but when I became a so-called cleaning expert, I was on less secure ground. The job was mainly about advising the manifold service and manufacturing entities within British Rail which cleaning agents and techniques were best suited to a particular problem. If a BR specified cleaner was not appropriate, then I had to approach the commercial sector. Naturally they were only too keen to be awarded a BR contract so after promoting their particular wares, cleaning company reps seemed duty bound to treat me to lunch. Lubricant suppliers were similarly disposed as indeed were pest control contractors. It was a hard life, but somebody had to do it!

Dave Smith

My early years in Normanton, Derby, during the war may have impacted on me, as did both my grandfathers. It was only latterly that I realised how they may have influenced my life. One was mischievous and the other quite a famous chemist at Steel Peach and Tozers in Sheffield. My exposure to them was mainly limited as we moved to Sheffield for only a few years after the war and then back to a more rural area near Littleover, Derby.

My maternal grandfather had an intoxicating effect. When he was charged with looking after me, when I was about two years old, he would ply me with tea which had a soporific effect. I was to learn later that milk was normally white and that Grandpa's milk was brown mainly because it was brandy. So, I learnt some chemistry from him — and I am still on the brown milk.

I moved from Sheffield aged about six and was absorbed into a gang of boys in Havenbaulk Avenue. They were a menace to the neighbourhood and well known for mischief-making. Calcium carbide was readily available for use in lamps in those days, but we found it was useful for making bombs. Calcium carbide mixed with water generates acetylene gas which is highly flammable. The fun we had with carbide bombs in people's gardens or nearby fields was enormous.

My Uncle Sid was also influential in providing me with chemicals and equipment; not just the standard chemistry kit for children of the age of about 14 or 15. Sid's included sodium metal, and my brother and I took great delight in placing pieces of this in my father's water butt. The explosions were memorable: just one small piece of sodium would cause a huge fountain of water to shoot up in the air, frightening the neighbours who had not forgotten the war and the sound of bombs and depth charges.

I went to a secondary school where, for most, the greatest expectations were to work on farms, clean the streets or, if you had talent, become an apprentice. I chose mischief-making as a career until the headmaster caned me three times in two weeks. He said he was going to win and perhaps I should use my talents better. I chose to take an exam and passed to go to college and get some GCEs, with chemistry and maths being my favourite subjects. At college I met John Hudson who worked at BR's Derby Calvert Street laboratory, who told me they would employ anyone. So, when I decided to continue my degree part-time, it came to pass that I joined, aged 19, an organisation of amazing people who gave me wonderful opportunities.

It was here that I became absorbed in the application of science for more beneficial purposes and hopefully to the benefit of the railway. Occasionally this necessitated me taking risks. I became involved in the writing of national standards at the British Standards Institute (BSI) in 1990 by accident (Jim Ward sent me to a meeting to represent BR). It was my pleasure to chair several committees, including those on environmental performance and occupational health and safety management systems. This led to me to representing the UK across the world, drafting standards for the International Standards Organisation (ISO) and the European body CEN.

It is perhaps surprising, considering my past as a Stink, that I should have had the honour of chairing the international committee (ISO) on occupational health and safety management that produced ISO 45001 in 2018. The irony of this appointment, considering my previous adventures, was pointed out by Sara Walton at BSI who gave me the impetus to produce this book. The standard will (hopefully) be adopted by hundreds of thousands of organisations trying to make the workplace safer, and it is hoped that 0.5-million organisations will have adopted it within ten years of publication.

Ian McEwen

I had no prior interest in railways, never went train spotting nor had a model train set, and I had probably travelled by train less than half a dozen times. However, in 1967, fresh out of university with a degree in Applied Chemistry, I joined British Rail Research.

I was interviewed in Hartley House by 'Tommy' Tomlinson and Dr Ian Dugdale, who was keen to follow up on the fact I had studied surface chemistry and found the new science of tribology (a multidisciplinary study of friction, lubrication and wear) very exciting. He explained that the department was about to expand significantly and would include a Tribology Unit. He also said that were I to be offered a post, there was a possibility the work might be suitable to register as an external PhD student with my chosen university. After the very friendly interview, I was shown around the Scientific Services laboratories and had a long chat with a certain David Smith, who was operating the infrared analysis equipment.

Some weeks later I was offered a post as scientific officer (grade II), with the suggestion I revisit Derby to discuss the idea of working towards a PhD and to meet Tony Collins, who would be in charge of setting up and recruiting for the new Tribology Unit. In August 1967 I became the first member of a team so new that the laboratories were not yet ready. Over the next six months, more people were recruited, including a senior scientific officer — none other than David Smith.

In my early years with BR Research, I worked on research into fundamentals of wheel/rail adhesion and on remedies for situations of low adhesion that could affect the traction or braking of rolling stock. The work included a good deal of time spent on the trackside as well as in the laboratory. My chemistry did not feature strongly, but in 1973 I was awarded the degree of PhD from the University of Salford for my thesis, Surface Chemical Effects In Wheel/Rail Adhesion.

By 1977 I was working on railway wheel and rail wear, particularly in curving situations, and was promoted to senior scientific officer grade in the Lubrication and Wear Unit of Scientific Services, taking my wear studies and equipment with me. In 1979 I moved into the Vehicle/Track Interaction Unit of Mechanical Engineering Research as principle scientific officer, a chemist among a team of engineers and mathematicians. I was responsible for practical and theoretical studies in vehicle curving behaviour and wheel/rail wear, including development of predictive techniques for use at the design stage. This included field studies and full-scale laboratory wear testing to validate theoretical predictions.

A departmental reorganisation led me into a short-lived spell in Civil Engineering Research's Track Mechanics Unit before I returned in 1986 to Scientific Services, this time as head of the Lubrication and Wear Unit. My remit was to manage a team carrying out research and ad hoc problem solving in a wide range of subjects under the general heading of tribology. Also included was specification of lubricants and fuels, technical investigations for, and advice to, all BR's functions and departments. I was responsible for coordinating the programme of condition monitoring by oil analysis of locomotives carried out by the regional laboratories.

A further reorganisation in March 1993 resulted in a change of title to Head of the Materials Science Unit, taking under my wing technical responsibility for battery technology, protective coatings for rolling stock and railway structures, quality audit and contract management. In 1994 soil mechanics, concrete and building materials as well as railway optics were added.

Although I started out with little interest in railways I soon got drawn in to the 'railway family' as I worked with and for a great many departments which make up the system. As well as work in the laboratory and office, I spent a great deal of time on and by the side of the track, in tunnels, on test vehicles, at maintenance depots and so on. For a great many years, I had a look-out certificate for track work; I was probably one of the highest qualified look-outs on BR.

The contributions I have made to this book involve very little chemistry but rather records the situations and activities a chemist on the railways found himself involved in.

Vince Morris

I had what (because I did not know any better) I assumed to be a normal education in the leafy Kentish suburbs: 11 Plus, O levels, A levels and then off to uni (which should be university, but no one calls it that now; too long a word to hold the concentration into the second syllable) in Canterbury (Kent, not New Zealand) to read chemistry. Not because it was local (I stayed in college or digs and home could have been a million miles away, rather than 40), but because it was new.

Ever since Z-Cars I had wanted to be a policeman, but I lacked one thing: guts. When I saw Marius Goring in The Expert (for the uninitiated, a forerunner to all the pulp forensic programmes now sprayed across a myriad of TV stations), I decided I would be a forensic scientist. Come the third year, uni very kindly arranged a jobs fair. No one from the Forensic Science Service, but the British Railways area manager (AM) from Canterbury made an appearance. I've always liked trains, ever since my father, who commuted daily to London, complained about the rotten service provided by the Southern Region. Being a contrarian, I studied the problem of transporting hundreds of thousands of commuters daily into London and delighted in its complexity. I even talked myself into the signal box at London Bridge and saw the effort required to signal two trains a minute through Borough Market Junction. I didn't care about train numbers or how many rivets were required to hold the locomotive together, it was the fact that it worked (sort of) which impressed me.

Back to the AM and the jobs fair. I thought I would seek an interview with the gentleman, not with the view of getting a job but out of interest, since I knew I could talk train operation to him. He was at least honest when he said he had not expected any scientists to make appointments, and he had no idea what scientists did but he knew that the Railway Technical Centre had just opened in Derby and he would arrange for me to go and see "the chemists there." And he did, God bless him. Meanwhile I had applied to the Forensic Science Service and an interview was arranged at their laboratory in Cardiff, which specialised in document examination, where there was a vacancy. I was even met by a police car at Cardiff station. The interview was a disaster, mainly because I was put off balance by the interviewer insisting I sat at the far end of the room from him, presumably so he could gauge my performance in the witness box. It is not the easiest thing to shout at the interviewer. I left despondent, but then I was invited to Derby, and even had a free ticket provided.

I expected an intimidating interview such as I had just experienced but no, there was no interview at all. Instead I was shown round the labs in Hartley House (not sure by whom) and eventually asked whether I had found it interesting. I replied positively (and truthfully) and was told that I may be contacted later, if a vacancy arose, to go for a formal interview at the location where such vacancy had arisen. It was in August 1969 I found myself having an interview at the London laboratory at Muswell Hill. My interviewer was Matt Pope, the Assistant Area Scientist. Dr Gordon Wyatt, the Area Scientist, was not available. Matt was the most mild of men but when we discussed the merits or otherwise of the Inter City branding (a bit too avant-garde for him), I was suddenly aware that I was simultaneously leaning back on my chair and banging the desk: not, I thought, a good move if you are trying to impress. Recovering from the shock, Matt mentioned the laboratory undertook forensic work for the British Transport Police, and I became even more concerned that I had blown this interview too. But a few days later, a letter came giving me a start date in September. I would be in the Building Materials Laboratory, managed by Stan Filipek, a Polish emigre who had, reportedly, walked across Europe to escape the Nazis.

I arrived as requested and was met by Fred Fawcett, the chief clerk, who was slightly surprised by my arrival until he remembered that a supernumerary graduate had been appointed on the recommendation of Matt Pope "because of his obvious and heartfelt interest in railways". Stan Filipek agreed to take me under his wing, as the vacancy which had caused me to be interviewed had been filled by another new starter. Fred Fawcett was an interesting character; a Desert Rat, of which he was justifiably proud. He was the fount of all knowledge on railway procedure and had the unerring ability of always having an empty desk at the end of the day.

A few months after I joined, Dr Wyatt retired and was succeeded by Alan Astles from Derby. It was then I came across Eric Henley, the Head of Scientific Services, for the first time. Eric had come to introduce Alan to the staff, but when he saw me his immediate question was "who are you?" in such an accusatory tone that I felt guilty and wondered whether my BR days were numbered. It was completely out of character and, as I soon learnt, Eric was the most gentlemanly of gentlemen. Maybe he was concerned that he had not recognised one of his own underlings. I do not know how many staff Alan subsequently spoke to, but he certainly called me into his office shortly after he took up his post and asked where I saw my future. I replied that I had had a long interest in forensic work, and so it was that I held a supernumerary post in the Building Materials Laboratory while undertaking forensic work under the watchful eye of Jean Ford, who did a large proportion of the police work. If I did not have a split personality before, I certainly had one by the time I had been with BR for a couple of years.

I stayed at London laboratory until, following the fire on the sleeper train at Taunton on 6th July 1978 in which 11 people died, it was suggested at the subsequent inquiry that British Rail should do something about fires on trains. The something it did was to set up a Fire Technology Team within Scientific Services, based in Derby. Thus it was that in 1981 (less than three years after the fire; pretty quick action for BR), I found myself heading for a new life at Derby HQ as team leader (based mainly on my experience of arson investigation for the BTP), after 11 or so years at Muswell Hill.

Vince sadly passed away before this book was published. He dedicated a lot of his time latterly, ensuring the document was edited and flowed. The authors all recognise the great contribution he made.

Ian Cotter

I was brought up in the Big Smoke when it was smoky! We had the pea-souper fogs in the 1950s caused by our own coal fires. My father was usually to be found in the evenings constructing model locomotives in our living room, with the soldering iron and vice in position as he manipulated small pieces of nickel-silver and brass. I think that must have given me a wish to get into engineering and a knowledge of railways. I admit to being one of those boys who frequented the large London stations and depots (on Sundays — fewer staff about!) with an Ian Allan book in hand and later the brownie camera for that wonky black and white shot.

I was not highly successful at school but with a wish to be an electrical engineer and a love of railways, I looked for a job in the rail industry. I discovered I had an aunt who was well connected with some railway engineers and found myself introduced to one who opened a door or two such that I was called to Derby for an interview. This resulted in an offer of a place on a training scheme which included a university degree in electrical engineering. However, my two E grades at A Level in chemistry and physics were nowhere near the three B grades needed for the course (especially the fail in maths — oh dear!). Those kindly engineers were also well connected and introduced me to Gordon Wyatt, the Area Scientist at the London laboratory, and I got a job there, starting on a fateful 29th September 1969, the same day as my good friend Vince Morris, now sadly deceased, also joined. Employment was in various departments in the lab, most of which I found fascinating. Main lab, under Len Bowyer, covered everything no one else did; rubbers, claims, cleaning materials, quality control, engine coolants and all sorts of oddities. The Southern Region looked upon us as their lab so there was frequent contact with engineers and depots in the south.

Day release for an HNC in chemistry started, immediately followed by a course to gain a Licentiateship of the Royal Institute of Chemistry, but I still hankered after being an engineer so persuaded the management (with the now Area Scientist Alan Astles's support, and the involvement of the trade union) to offer me a thin sandwich course (six months at uni and six months' work) in chemistry. With a year off for good behaviour (or the HNC to be exact), I qualified in 1974 with a BSc in Applied Chemistry. I then looked for a route into engineering — nothing if not persistent! Most engineering institutions demanded a design project (a skill I did not possess) but one did not: the Institute of Energy (IE). Thus, it was after a year of three evenings a week at Borough Polytechnic that I qualified in fuel technology and combustion engineering and joined the IE, gaining full membership and chartered engineer status in 1981. Meanwhile, I had been accepted for professional membership of the Royal Institute of Chemistry (now the Royal Society of Chemistry) gaining Chartered Chemist and Fellowship in 1991.

I had been around the various sections at the laboratory and ended up in oil analysis when high speed trains (HST) were coming into service. Liaison with engineers ramped up rapidly as engine failures multiplied and a one or two-week turnaround of sample results became three hours, with a much larger team successfully supporting the iconic HST by preventing most incipient faults developing into a wrecked engine. Elevation to Assistant Regional Scientist in London on the retirement of Matt Pope (he held my great respect as a knowledgeable scientist who would help anyone with technical queries) ensued, followed by a successful application for the Head of Laboratory (then known as Regional Scientist) at Swindon. I had not realised that Swindon lab was on the cusp of a severe staff reduction, as Swindon Works was slated for closure, but with the help of lucrative external work for a well-known car manufacturer (again with a three-hour turnaround) the lab staff numbers began to increase again. Closure of the London lab (which had been taken over from BR Research by Network South East) in 1993 allowed increased sales activity and most of the Southern Region and some of East Anglia was added to the Swindon portfolio. Privatisation was now on the agenda, but thereby hangs another tale!

Geoff Hunt

I was always fascinated by the way that sparklers behaved on Bonfire Night. Just how could a piece of dull grey metal give such a display? Hence my early, innocent interest in chemistry was established. The next enlightened moment came when, around 10 years of age, a No 1 Chemistry Set appeared at Christmas. Now I had the real thing to play with. Early experiments were not too successful, heating up log wood chips in a test tube over a methylated spirit burner on my mum's dining room table taught me a lesson (follow the instructions) and do not put a cork in the top of the test tube being heated. The coloured stain on the dining room ceiling lasted some time as a reminder!

Teaching chemistry in secondary modern schools in the 1960s was not mainstream but I was lucky to have a science teacher, Ken Redfern, who must have seen something in me. His encouragement to go after O Levels when I left school placed me firmly on the chemistry ladder. Unfortunately, full-time A Levels were not to be and my dad's instruction to find work was the next challenge. An initial position as a trainee metallurgist at Qualcast Foundries Ltd provided an opening to the chemical laboratory. I had learnt to follow recipes during my short spell making metal and this led me to my first post as a 'bench monkey' in the foundry chemical laboratory. It also opened up the opportunity for further education in chemistry, part-time (a day and two nights, sometimes three nights, 12 hours in the classroom sometimes followed by a night shift as the lone working foundry chemist on site — obviously never allowed today under health and safety rules!), for nine years leading eventually to membership of the Royal Society of Chemistry and Chartered Chemist status. Along the way a further 'bench monkey' experience continued with quality control analysis of blended and reclaimed synthetic aviation gas turbine oil and hydraulic system oils at Dalton and Son Ltd, Silkolene Oil Refinery in Belper.

April 1st 1974, and I arrived at Hartley House laboratory, Derby, for my first post as a technical officer. As part of the probationary period, all new starters began their railway chemist career on the oil bench under the tutelage of John Sheldon, whose primary claim to fame was his ability to have completed the Daily Telegraph cryptic crossword before he arrived for work each morning. Having just come from an oil refinery background, kinetic viscosity measurements, water content and solids analysis were second nature and probation period successfully completed, I began my career as a railway chemist in Scientific Services. The oil bench also opened up the opportunity for effluent sampling and analysis which led to visits to railway depots with strange names like Beal Street, Bescot, Soho, Saltley and Oxley across the Midland Region.

As I became more experienced, I was given a Hartley House aisle laboratory bench, which was a kind of official recognition of achieving formal railway chemist status, albeit still learning. Further experience was gained carrying out ad hoc analysis in accordance with the laboratory 'Black Book' volumes, plus effluent and water sampling and analysis, microbiology, cleaning technology, unknown substance identification... the opportunities were endless. In the downstairs Hartley House laboratory, my training and education continued with instrumentation chemistry carried out under Brian Littlewood and his intriguingly named number two, Heinz Bauer: X-Ray diffraction, spectroscopic metal analysis on wonderful sounding machines called the Quanta-met and Quanta-vac. It was here I encountered infrared analysis, gas chromatography, nuclear magnetic resonance instrumentation... all of which supported my continuing college education.

The defining career change came around 1980, room 106 in the upstairs of Hartley House laboratory where white-haired Phil Dimmick was the Derby occupational hygienist, the primary work area being atmospheric asbestos sampling and analysis and awareness training for railway staff. And so a new career in occupational hygiene began. More college training followed for Preliminary Certificates leading to Licentiate qualification of the British Occupational Hygiene Society covering all aspects of the discipline; monitoring noise, harmful gases and vapours, the thermal environment, asbestos of course, and many other specialist subject areas that affected the railway workplace. The 1980s and 1990s proved defining years in the subject, European legislation brought about a raft of UK laws covering asbestos, the Control of Substances Hazardous to Health [COSHH], Noise at Work to name but a few — though all relevant to and impacting on the railway environment.

Occupational hygiene support to BR over this period increased significantly, not just in Derby but across all of Scientific Services laboratories, with three or four staff either fully or partially trained in the discipline under the technical direction of senior occupational hygienist Lou Hirsch, based at Muswell Hill. The 1980s was also the time of some historical rail industry disasters, the Summit Tunnel fire and more importantly the King's Cross Underground fire. It was then that the various health and safety skills within in Scientific Services were brought together under Vince Morris in the Health and Safety Unit, managed from Derby, including responsibility for the wider occupational hygiene service. Environmental management came to the fore and so the old adage of 'dilution was the solution to pollution' no longer applied. The principles of control and elimination became increasingly important and personal protective equipment only to be used as a final resort.

In the 1990s, privatisation loomed. Scientific Services was being primed for sale, so we became commercial, selling our services outside of the railway industry and Business Safety Manager was added to my CV in 1992. December 1996 saw the end: Scientific Services, by now renamed Scientifics, was being sold and 132 years of real railway chemistry history ended.

A new career beckoned as a safety management consultant and sector company head of safety for a global engineering design consultancy, but it was my railway chemist heritage that defined me.